Design and Appraisal
of Hydraulic Fractures

Design and Appraisal of Hydraulic Fractures

Jack Jones
BP Canada Energy Company

Larry K. Britt
NSI Technologies

Society of Petroleum Engineers

ISBN 978-1-55563-143-7

12 13 14 15 / 9 8 7 6 5 4 3 2

Society of Petroleum Engineers
222 Palisades Creek Drive
Richardson, TX 75080-2040 USA

http://www.spe.org/store
service@spe.org
1.972.952.9393

Introduction

This book is a brief introduction to technical aspects of hydraulic fracture completions. The book is organized around the main steps of job design, fracture placement, evaluation of post-completion results, and performance prediction. As such, the book presentation tracks what is essentially a standard "plan + do + measure + learn" completion-evaluation cycle as applied to hydraulically fractured wells. Chapter 1 covers job design and execution/placement, which reflects the "plan + do" parts of the cycle. Chapter 2 is devoted to the evaluation and comparison of the actual in-situ stimulation with that designed, while Chapter 3 outlines the evaluation and prediction of subsequent well performance. These last two chapters reflect the "measure + learn" cycle elements.

The book is aimed at a broad audience of petro-technical professionals, geoscientists, and engineers of all disciplines. No prior experience in the technical areas or methods discussed is assumed or required. We have tried to structure the book for easy reading. Each subsection includes a boxed summary of the key points and methods presented; reviewing these summaries should help the reader gain an understanding of the most important concepts in each subsection and how that material fits into the overall flow of the book.

As an overview of the state of the art, the book obviously weaves together and compiles the documented contributions of many researchers and practitioners. As such, our main contribution is in the interpretation presented of these original contributions and the discussion of our experience with their use in answering day-to-day engineering questions. For those who want to delve more deeply into a given subject, references are provided throughout to the primary sources describing the methods, approaches, recommendations, and conclusions of these original contributors.

A note on the figures and illustrations is probably warranted. Whenever possible, the figures and illustrations used here have been taken directly from the original references. When this was not possible or appropriate, we have created visuals ourselves to convey the information.

As with all publications, we have striven to avoid errors and incorrect statements. Any that remain are solely our responsibility. We hope that we have the opportunity to correct any such problems in future editions.

Acknowledgments

In any undertaking like this, help comes from many sources and ranges from context discussions to detailed rewriting. Though we are obviously thankful for all the help we have received, some contributions have been absolutely critical. In this category, we want to acknowledge Raj Raghavan and Hossein Kazemi for their substantial editing, insightful comments, and patient support. This book is, as are we, very much a product of the experience we both gained from our time working with colleagues at the Amoco Research Center. In particular, we would like to thank Mike Smith, Ken Nolte, and Ram Agarwal for all they have taught us. The patient guidance and support we have received from Jennifer Wegman at SPE has been invaluable, and we both thank her very much. Finally, we both would like to thank our employers, NSI (Britt) and bp Americas (Jones), for allowing us the space to complete this project.

Contents

Chapter 1

Design and Placement

1.1 Basic Introduction and Description

1.1.1 Process and Analysis History. Hydraulic fracturing has made a significant contribution to the oil and gas industry as a primary means of increasing hydrocarbon production. Since fracturing was introduced by Stanolind (Amoco) in 1947, millions of fracture treatments have been performed. Fracture stimulation treatments not only increase production rates but also are credited for adding billions of barrels of oil and trillions of cubic feet of gas not otherwise economic to develop. In addition, hydraulic fracturing has accelerated hydrocarbon production and significantly increased its present value. Today, hydraulic fracturing is applied throughout the world. In 1991, a French Petroleum Institute survey reported that 71% of all wells completed worldwide are fracture stimulated. For most operators, the second largest expenditure is well stimulation. The first is drilling, of course, and only rotary drilling technology has had as big an impact on hydrocarbon recovery as hydraulic fracturing.

Shortly after the introduction of hydraulic fracturing, the importance of fracturing pressure data was recognized by Godbey and Hodges (1958). Later, fracturing pressure and

---

Section Takeaway

- The history of hydraulic fracturing is long and successful.
- Hydraulic fracturing is a multidisciplinary process—engineering, geotechnical, and operational.
- The magnitude and direction of the principal stresses are critical to hydraulic fracturing.
- A fracture stimulation design is a three-stage process: the pad stage (no proppant), the slurry stage (with proppant), and the flush (slurry to perforations).
- The combined analysis of bottomhole pressure and microseismic images could provide a unique framework for understanding the hydraulic fracturing process.
- Slurry dehydration or leakoff occurs as the fracturing fluid filtrate invades the reservoir pores and creates a filter cake at the fracture face.

its relation to in-situ stresses were included in the pioneering models of Khristianovic and Zheltov (1955), Perkins and Kern (1961), and Geertsma and de Klerk (1969). However, several more years passed before the analysis of fracturing pressure data started to become an accepted industry practice.

In 1978, Amoco Production Company initiated a coordinated program of field data collection and analysis to improve the understanding of the mechanics of the fracturing process (Veatch and Crowell 1982). A series of papers from the 1979 SPE Annual Technical Conference and Exhibition presented results from this program, including a paper by Nolte and Smith (1981) that introduced a scientific basis for the interpretation of pressure behavior during fracture treatment, and one by Nolte (1979) for interpreting pressure decline after the treatment. Additional developments include the use of long-spaced-sonic logs to determine reservoir stress and post-fracture diagnostic temperature (Dutton et al. 1982) and tracer logs to determine fracture height (Dobkins 1981).

Nolte (1979) presented procedures for quantifying fluid-loss coefficient, fracture length and width, fluid efficiency, and time for the fracture to close from a "minifrac" test procedure which could then be used in designing the actual fracture treatment (Schlottman et al. 1981; Elbert et al. 1984; Dobkins 1981; Smith 1985; Smith et al. 1987; Morris and Sinclair 1984). The work by Nolte and Smith was extended to include analysis for determining proppant and fluid schedules from the fluid efficiency when little or no information is available (Nolte 1986a). In addition, theoretical work covered three popular 2D fracture geometry models (Nolte 1986b), more-complex geometries involving fracture height growth (Martins and Harper 1985), and pressure-dependent fluid loss (Castillo 1987). In fact, service companies routinely perform mobile on-site collection and analysis of fracturing pressure data using Nolte and Smith's (1981) techniques.

Hydraulic fracturing of oil and gas formations has a long history. Oil-well fracturing, for example, can be traced to well-shooting with liquid nitroglycerin in the late 1800s in the northeastern United States. Hydraulic fracturing with acid to stimulate well performance was introduced in the 1930s. The idea of hydraulically fracturing a formation to enhance the production of oil and gas was first conceived and advanced by Floyd Farris of Stanolind Oil and Gas Corporation after an extensive study of the pressures encountered while squeezing cement into formations (Farris 1953). The first attempt to hydraulically fracture a well for stimulation was conducted in the Hugoton gas field in Grant County, Kansas, in 1947. A total of 1,000 gallons of napalm-thickened gasoline was injected, followed by a gel breaker, to stimulate the Chase formation in the Hugoton field at 2,400 ft. No proppant was used in this first treatment, and the deliverability of the well was not changed appreciably. Interestingly, the industry found out early on that the use of proppants, even in small amounts, was critical to hydrocarbon recovery. The hydraulic fracturing process was first introduced to the industry by J.B. Clark in 1948 (and patented and licensed by Clark in 1949).

Hydraulic fracturing consists of initiating a fracture in the formation with the hydraulic pressure of a fracturing fluid, propagating the fracture with fluid, and holding the created fracture open with proppant. The propped fracture becomes the conductivity pathway between the formation and the wellbore for hydrocarbon production. To serve this pathway function, a fracture stimulation design has three principal stages (**Fig. 1.1**): the pad stage, the slurry stage, and the flush stage. The pad stage (without proppant) is used to initiate and propagate the fracture, develop fracture width, and provide sacrificial fluid for leakoff. The slurry stage (with both fluid and proppant) is used to position the proppant in

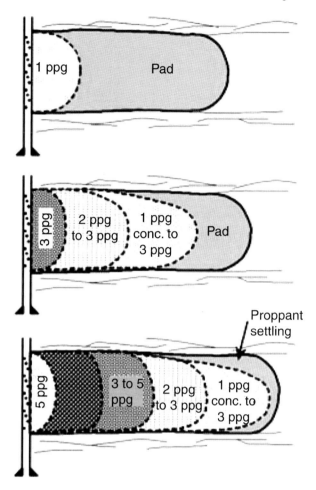

Fig. 1.1—Proppant transport and the optimum fracture design.

the fracture so that there is a constant proppant concentration through the length of the fracture at the end of pumping. This is achieved by ramping up the proppant concentration to the desired maximum concentration. In the process of *slurry dehydration* (*leakoff*), the lower proppant concentrations that have progressed toward the tip of the fracture have dehydrated in the fracture such that the resulting proppant concentration in the fracture equals the final maximum proppant concentration. Finally, the treatment is flushed to the perforations. Although the fracture treatment can be designed for optimal well performance, the success of the process is outside the control of the design/execution engineer. For example, the direction of the fracture propagation is controlled by the state of the in-situ stress. The fracture geometry is similarly controlled by geomechanical parameters and leakoff.

The three principal stresses that control fracture geometry and propagation are the vertical stress, the minimum horizontal stress, and the maximum horizontal stress **(Fig. 1.2)**. Most often, the maximum principal stress is the vertical stress—often equal to the weight

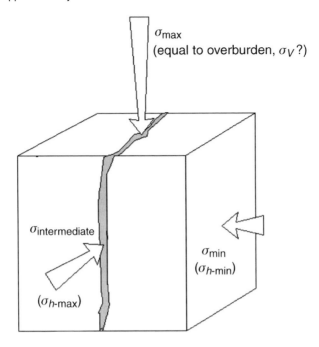

Fig. 1.2—General stress state.

of the overburden—while the maximum horizontal stress is generally the intermediate principal stress. In this normal faulting environment, the hydraulic fracture propagates as a vertical fracture in the direction of the intermediate stress (the maximum horizontal stress). Because the direction of the fracture is controlled by the in-situ stress state, hydraulic fracturing generally is not a good way to establish communication with a desired reservoir target hundreds or thousands of feet from the wellbore.

Although the stress state identified in Fig. 1.2 is fairly typical, it by no means represents all of the possibilities. For example, at shallower depths, where the weight of the overburden is small and the vertical stress is not the maximum principal stress, a horizontal—rather than a vertical—fracture propagates. And in some reservoirs at depth, where the principal stresses have been altered by geologic activity such as salt or shale intrusions, horizontal fractures may result.

Perforation design is also extremely important to successful fracturing. For example, if the in-situ stress state favors a preferred fracture orientation, misaligned perforations may result in extremely high treating pressures and poor wellbore-fracture communication, which could jeopardize the fracture execution and achievement of design objectives. If, on the other hand, the minimum and maximum horizontal stresses are similar, such issues are far less important. For example, in the Gulf of Mexico, the horizontal stresses are generally similar, and few problems are encountered in fracturing deviated wells, while in the North Sea, where the horizontal stresses can differ significantly, many problems occur, and S-shaped wells are often drilled (i.e., vertically through the reservoir) to alleviate some of the relevant issues. **Fig. 1.3** shows a plot of the instantaneous shut-in pressure (ISIP) fracture gradient (ISIP/depth) vs. well deviation for a field in the Norwegian sector of the

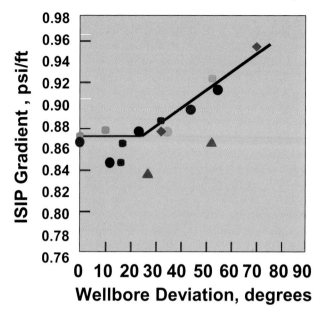

Fig. 1.3—Effect of well deviation on the fracture gradient.

North Sea. As the well deviation increased, the pressure required to pump a fracture stimulation increased. Once recognized, the well deviation was aligned with the maximum horizontal stress (fracture direction), and the pumping problems were alleviated. Finally, hydraulic fracturing involves moving a (sometimes large) portable chemical plant to the well location.

Halliburton Oil Well Cementing Company was given an exclusive license on the new process. The first two commercial fracturing treatments were performed in Stephens County, Oklahoma, and Archer County, Texas, on 17 March 1949, using lease crude oil or a blend of crude and gasoline and approximately 100 to 150 pounds of sand. Both wells were successful and, thereafter, application of the fracturing process grew rapidly (Maly and Morton 1951). Note that, in the early days of fracturing, all of the created fractures were thought to be oriented horizontally (Clark et al. 1953; Huitt and McGlothin 1958; Huitt et al. 1958; Huitt 1960). Hubbert and Willis (1957) used a gelatin model to show that nearly all fractures were vertical, and Anderson and Stahl's results (1967) confirmed that most fractures were indeed vertical. The first fracture stimulation to use more than one-half million pounds of proppant was performed by Pan American Petroleum Corporation in Stephens County, Oklahoma, in October 1968. Today, fracture treatments are performed regularly all over the world (**Figs. 1.4 and 1.5**).

There are a number of reasons to conduct hydraulic fracture stimulations, including damage mitigation, production enhancement, and formation control. Significant technical advancements have been made in the 50-plus years since the first commercial treatments. The first few commercial fracture stimulations consisted of small fluid and proppant volumes and were designed to bypass near-wellbore damage. In the late 1970s and 1980s, massive hydraulic fracture stimulations were conducted to improve the production performance and reserves recovery of tight gas formations. In the early 1980s, tip-screenout

Fig. 1.4—Land-based fracture stimulation. Courtesy of BJ Services.

Fig. 1.5—Offshore fracture stimulation. Courtesy of BJ Services.

(TSO) fracturing was developed and used to increase fracture conductivity and well performance. In the 1990s, frac packing—a combination of TSO fracturing and gravel packing—was used for formation control in high-permeability unconsolidated formations. Cleaner and more suitable fluid systems (gellants, crosslinkers, and breakers) were developed. Higher-strength synthetic proppants were introduced for deep-well applications.

Hydraulic fracturing is complicated by the fact that the events cannot be directly observed. This complication has been overcome by indirect analyses based on wellbore pressure and flow rate. In addition, with the development of microseismic monitoring, fracture dimensions can be determined by coupling pressure analysis to microseismic mapping.

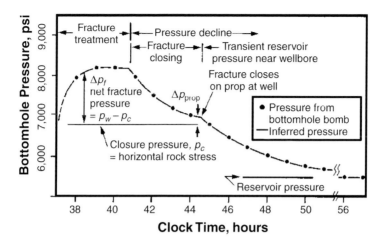

Fig. 1.6—One of the first published treating pressure and pressure decline records.

Analysis of fracturing pressure response is analogous to pressure transient analysis in reservoir engineering. However, pressure analysis of reservoirs is a mature discipline, while its application to hydraulic fracturing is still evolving. **Fig. 1.6** shows one of the first published recordings of bottomhole pressure during and after a fracture treatment. The analogy to transient pressures in reservoirs can be seen in the figure, with increasing pressure during injection and the pressure falloff or decline after shutdown. The figure also shows that during the first half of the treatment, the pressure was increasing, while during the last half of the treatment, the pressure remained essentially constant. The analysis of fracturing pressure has become more definitive with the addition of microseismic fracture mapping. Microseismic imaging is more difficult and costly to measure than the treating pressure. However, occasional use of microseismic imaging, coupled with pressure analysis, should be sufficient to develop a geomechanical model of an area.

1.2 The Fracturing Process and Its Elements

1.2.1 Pretreatment Planning. Prefracturing planning encompasses data collection, treatment design and documentation, and interaction between the service provider and the operator. Data include reservoir parameters necessary to estimate well potential and geomechanical data for fracture geometry estimation. In practice, it may not be possible or economical to collect data from every desirable source. However, in an initial development well, an effort should be made to fully understand the well from all perspectives.

To justify fracture stimulation, the formation flow potential from fracturing must be critically evaluated. The reservoir parameters required for performance estimation include porosity, water saturation, permeability, and reservoir pressure. These parameters can be determined from logs, cores, and pressure-transient testing. Other information that can aid the evaluation of well potential includes an understanding of the mineral constituents of the formation; the proximity of gas, oil, and water contacts; and an estimate of drainage area.

Once it has been established that a fracture stimulation provides economic viability, certain geomechanical data are required to estimate the fracture dimensions and the

> ### Section Takeaway
> - Prefracturing planning should include interaction with all stakeholders.
> - Prefracturing data should be collected to design a fracture treatment.
> - Perforation hole size, phasing, and interval length should all be considered preparatory to fracture stimulating.
> - Perforation hole size should be at least 7 times the maximum proppant diameter.
> - Perforations should be phased at 60 or 120°.

preliminary treatment design required to optimize oil and gas recovery. The geomechanical data include the minimum horizontal stress, Young's modulus, and leakoff coefficient. The minimum horizontal stress and stress contrast can be estimated with Hubbert and Willis' or Eaton's equations, use of long-spaced or dipole sonic data, and/or review of offset fracture stimulation data, but should be validated with prefrac injection testing. Young's modulus can be estimated from sonic logs (dynamic Young's modulus from logs is generally 30 to 50% greater than static Young's modulus) or prefrac injection testing (nonunique, because it is coupled to the leakoff coefficient), or measured by triaxial compression testing of formation core (the recommended method). The leakoff coefficient can be estimated by calculations based on fluid, wall-building characteristics, and/or offset well-fracture stimulation data. Spurt loss (filtrate lost to the formation prior to the development of a filter cake) can be determined from a core study. Other data include reservoir temperature, overburden stress, fluid injection rate, and limitations associated with wellbore tubulars, fluid rheology, and proppant density.

Injection/decline tests and minifracs may be required on only a select few wells early in the development program to determine formation stresses and leakoff. Also, overburden stress and Young's modulus should be required on only early wells, unless the geology varies significantly from one area to another in the development region or the cost of stimulation failure is well in excess of the costs of testing, as in many offshore environments.

A preliminary treatment design should be formulated to aid in pretreatment planning. While this might not be the final design, this will provide estimates of treatment requirements, including fluid, chemical, and proppant amounts; on-site storage; fracturing equipment;

> ### Section Takeaway
>
> *Hubbert and Willis Equation*
> $$\sigma_{h\min} = p_{frac} = K * (\sigma_{ob}-p) + p + p_{tec}$$
>
> *Eaton's Modified Equation*
> $$\sigma_{h\min} = p_{frac} = (v/1-v) * (\sigma_{ob} -p) + p + p_{tec}$$
>
> where $\sigma_{h\min}$ is the minimum horizontal stress (fracture pressure), v is the Poisson's ratio, σ_{ob} is the overburden pressure, p is the reservoir pressure, and p_{tec} is the tectonic component.

TABLE 1.1—PERTINENT DATA REQUIREMENTS FOR PROPER PREFRAC PLANNING		
Category	Parameter	Method
Formation	Porosity	Core and logs
	Water saturation	Logs and core
	Permeability	PTA, core, and logs
	Mineralogy	Core
	Reservoir pressure	Pressure transient test
Wellbore configuration	Casing/tubing	Depth, size, and weight
	Packer	Depth and type
	TD/PBTD	Depth
	Perforations	Size, density, and phasing
Fracture geometry	Minimum stress	Prefrac injection tests
	Young's modulus	Core
	Overburden stress	Density log
	Reservoir temperature	Log/static measurement
	Leakoff coefficient	Prefrac injection tests
Fracturing materials	Compatibility (formation)	Core tests
	Compatibility (fluids)	Core and fluid tests
	Proppant transport	Fluid viscosity
	Fracture conductivity	Size, type, and concentration
Preliminary frac design	Fluid and proppant volume	
	On-site storage	
	Equipment	
	Location size	
	Personnel requirements	

location, sizing, and equipment placement logistics; and personnel requirements. The expected treatment schedule should be reviewed with the service company at the earliest possible date to ensure that all requirements can be met.

Prefrac planning and execution should be prepared either by the operator or jointly by the operator and service provider for each treatment and should include all of the pertinent wellbore, completion, fracture stimulation, and reservoir data. Prior to the fracture stimulation, these data should be reviewed by the appropriate stakeholders as part of the preliminary fracture design process. Such data review is a critical part of the pretreatment planning to minimize subsequent operational issues. **Table 1.1** summarizes a checklist of the pertinent information required for prefrac planning.

1.2.2 Surface Equipment and Well Preparation. To successfully conduct a fracture stimulation, the equivalent of a small portable chemical plant is required on location. Because of the large amounts of materials and equipment that are used to execute a fracture stimulation and achieve the treatment objectives, logistical planning and wellsite preparation can be extremely important. The equipment needed for a fracture treatment includes frac tanks, proppant silos, and additive bins for the storage of the base fracturing fluid, proppant, and additives, respectively. In addition, a blender is needed to mix the various components of the treatment, and pump trucks are required to pump the fracturing slurry through the wellhead isolation tool and down the wellbore with sufficient hydraulic horsepower (HHP) to initiate and propagate the fracture. Fig. 1.4 shows the massive requirements for pump trucks, fluid storage, and monitoring equipment.

1.2.3 Completion Considerations. Proper design of a perforating or completion program can have a tremendous influence on the success of a fracture treatment. In perforating, the perforating location, size, and orientation with respect to the fracture plane, as well as the number of perforations, affect how easily the fracture is generated and proppant is placed. For vertical wells, it is best to perforate most of the pay; in deviated wells, clustered perforations may be helpful to control the point of fracture initiation.

The perforation hole size is important because the hole entrance has to be large enough to readily pass the proppant. Hole size should be over seven times the largest proppant size (Gruesbeck and Collins 1982). Thus, for 20/40 mesh Hickory sand (maximum diameter 0.03 in. or 0.76 mm), the minimum hole size would be 0.21 in. (5.3 mm). This criterion puts the hole beyond the typical limits of proppant bridging in a crack or hole. In most cases, the hole diameter is insignificant once the diameter is larger than 0.5 in. Information on recommended perforation size for a specific proppant is available from the manufacturer.

The number of holes also plays an important role in determining perforation friction pressure, HHP requirements, and, ultimately, treatment costs. Perforation friction pressure (i.e., pressure drop across perforation) can be calculated by

$$\Delta p_{pf} = (0.2369) \, (q^2 \rho)/(N^2 d^4 C_p^2), \quad \dots\dots\dots\dots\dots\dots\dots\dots\dots\dots\dots\dots\dots \quad (1.1)$$

where ρ = fluid density in lbm/gal, q = pump rate in bbl/min, C_p = perforation discharge coefficient $0.3 < C_p < 0.95$, N = number of perforations open, d = perforation diameter, in., and Δp_{pf} = pressure drop across perforations in psi.

Eq. 1.1 shows that if one perforates a small interval (20 ft) through tubing ($d = 0.20$ in.) with 1 shot per foot (spf) for many fracture treatments, perforation friction pressure of 1,415 psi can be generated. This assumes a pump rate of 30 bbl/min, a fluid density of 8.4 lbm/gal, and a discharge coefficient of 0.625; and all perforations are open. Eq. 1.2 describes the HHP requirements:

$$\text{HHP} = 0.0245 q p_w. \quad \dots\dots\dots\dots\dots\dots\dots\dots\dots\dots\dots\dots\dots\dots\dots \quad (1.2)$$

One can see that perforating a small interval could require additional horsepower and put the fracture stimulation at risk.

In addition to the number and size of the open perforations, the orientation of the holes with respect to the orientation of the fracture plane is also of significant importance. BP and Arco (Martins et al. 1992) determined that orienting the phasing of the perforating gun to the plane of fracture growth decreases the chances of screenouts. Pearson et al. (1992) have shown that breakdown pressures are greatly increased when the perforations are not aligned with the fracture orientation. If the fracture plane is known and the guns can be reliably oriented, 180° phasing would be optimal.

Abass et al. (1994) show that perforation phasing should ideally be designed to ensure adequate numbers of perforations within 30° of the maximum horizontal stress. Guns oriented at 60° or 120° are thought to be better for stimulations because the perforation "banks" are closer to the plane of the fracture direction. Guns with 0 or 90° phasing should be avoided.

Perforation misalignment and reduced shot density put the treatment at risk because of the likelihood of a screenout, but they also increase the cost of the treatment because much higher treating pressures are required. In many areas, 1 spf and 0–180° phasing is standard for wells. In short intervals—less than 50 ft—it is recommended that at least 4 spf be used at 60° or 120° phasing to ensure that an adequate number of holes are aligned with the fracture direction (maximum horizontal stress).

The HHP requirements of improperly aligned perforations, coupled with the risk of screenout, make the cost of a few extra perforations insignificant. Deep-penetrating charges in phased hollow-carrier casing guns, conveyed by either wireline or tubing, are generally thought to create the best perforations for fracturing.

1.3 Fracture Design

1.3.1 Fracture Design Objectives.
The primary objective of hydraulic fracturing is to create and maintain a stable fracture with excellent conductivity to maximize well productivity and ultimate recovery. To appreciate the reasons, one must understand the relationship between the reservoir and fracture variables of permeability, fracture half-length, and fracture conductivity. The interdependence of these variables is best described by the dimensionless fracture conductivity, F_{CD}. The equation is

$$F_{CD} = k_f w / (k x_f) \qquad\qquad\qquad (1.3)$$

Here, k is the formation permeability, k_f is the permeability of the fracture, w is the fracture width, and x_f is the fracture half-length. Eq. 1.3 relates the fracture's ability to flow fluids to the wellbore with the reservoir's ability to flow fluids to the fracture. The parameters used to define F_{CD} are illustrated in **Fig. 1.7.**

- Fracture half-length, x_f, ft
- Formation permeability, k, md
- Fracture flow capacity, $k_f w$, md-ft

$$F_{CD} = \frac{k_f w}{k \cdot x_f}$$

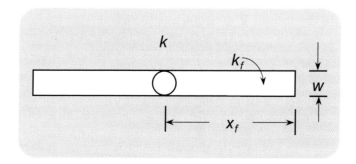

Fig. 1.7—Definition of dimensionless fracture capacity, F_{CD}.

Section Takeaway

- Well performance can be maximized in permeable reservoirs by achieving a dimensionless fracture conductivity, F_{CD}, of 2.3.
- In tight gas reservoirs, fracture length and not conductivity is the critical fracture parameter.
- Prefracture determination of reservoir permeability greatly improves fracture design and achievement of optimal fracture dimensions.
- Well spacing and drainage area play an important role in determining the optimum fracture dimensions.

In 1961, Prats showed that for wells in moderate- to high-permeability reservoirs, pseudoradial flow can be modeled by radial flow behavior. The use of radial flow allows modeling of the well behavior through the effective-wellbore-radius concept, r'_w. This concept provides a measure of the apparent increase in the wellbore radius as a result of a fracture. A fractured well appears to have a diameter that is larger than the true diameter. For example, from **Fig. 1.8,** we see that if F_{CD} is large, then $r'_w \approx x_f/2$; that is, large-conductivity fractures result in wells with an effective radius approximately equal to one-half the fracture length. In essence, fracturing wells in moderate- to high-permeability reservoirs is equivalent to expanding the wellbore of the wells. The effective-wellbore-radius concept is a powerful aid to understanding fractured-well performance in permeable reservoirs.

In lower-permeability reservoirs where transient flow is long (e.g., in a tight gas formation), this concept is less useful. Fig. 1.8 plots the ratio of effective wellbore radius to fracture half-length as a function of dimensionless fracture conductivity. This figure highlights the relationship between fracture length, conductivity, and reservoir permeability. If,

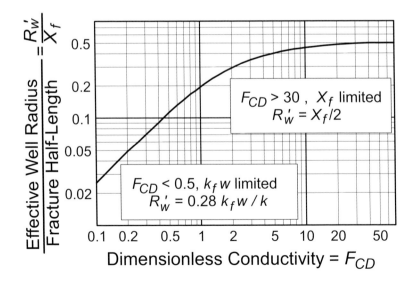

Fig. 1.8—Effective wellbore radius, r'_w/x_f, vs. F_{CD} (after Prats 1961).

for example, F_{CD} is low ($F_{CD} < 0.5$) the fracture has finite conductivity, and the reservoir fluids will tend to flow toward the wellbore rather than the fracture tip, the desired target. The figure also indicates that increasing fracture length would not result in improved reservoir response. Conversely, if F_{CD} is high ($F_{CD} > 500$), the fracture has infinite conductivity, so increasing fracture conductivity would not improve reservoir response. For practical purposes, fractures having $F_{CD} > 20$ act as infinite conductivity fractures, where the effective wellbore radius, r'_w, is one-half the fracture half-length. Prats' 1961 work took the wellbore analog further to develop the steady-state folds of increase (FOI) concept, which relates post-fracture well performance to prefracture performance.

$$\text{FOI} = \ln(r_e/r_w)/\ln(r_e/r'_w) \quad \dots\dots\dots\dots\dots\dots\dots\dots\dots\dots\dots\dots\dots\dots \quad (1.4)$$

This concept is used to show the benefits of hydraulic fracturing and is most applicable when the transient flow period is short. In addition, Prats' work (1961) showed that for any proppant volume, the productivity increase is maximized if the fracture is created and the proppant is placed to achieve an F_{CD} of 2.3. **Fig. 1.9** shows a plot of FOI vs. dimensionless fracture conductivity made from steady-state stimulations of constant volume fractures. Whether the stimulated reservoir is a 0.5-md or a 10-md reservoir, the maximum post-fracture productivity occurs for a dimensionless fracture conductivity of 2.3.

Thus, if the fracturing objective is to maximize well productivity, it can be accomplished for an F_{CD} of this magnitude. At the wellsite, this work can be used to maximize well performance. For example, when on location and redesigning a fracture treatment after obtaining the geomechanical and material balance parameters from a minifrac test, the redesigned treatment should have a minimum target F_{CD} of 2.3. This discussion highlights a fundamental fact of hydraulic fracturing: one of the most important parameters to achieving fracturing objectives is reservoir permeability.

Hydraulic fracturing is principally a rate acceleration process for single-phase flow (such as gas reservoirs); however, reserves growth can occur from a hydraulic fracture by improving the vertical communication of hydrocarbons with the wellbore, improving recovery from lower-permeability layers, and extending operating life. Hydraulic-fracturing objectives are

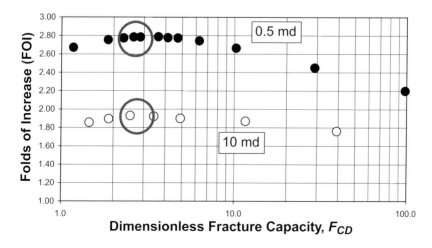

Fig. 1.9—Effective wellbore radius, r'_w/x_f, vs. F_{CD}.

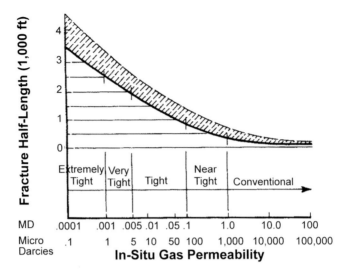

Fig. 1.10—Desired fracture half-lengths for different formation permeabilities. From Veatch (1983a).

best achieved through economic optimization, since nearly every fracturing decision (e.g., fracturing fluid; proppant type, size, and concentration; pump rate) is an economic one. There are numerous economic and well-performance criteria that can be used to optimize fracture stimulations. The most common are net present value, discounted return on investment, payout, initial potential, and annualized well performance (Veatch 1983a, 1983b; Britt et al. 1985). However, each of these parameters provides different optimum dimensions; therefore, the optimization criteria should be determined by corporate objectives.

Fig. 1.10 plots fracture half-length vs. reservoir permeability, showing that the optimum fracture half-length is inversely related to permeability. That is, as reservoir permeability becomes smaller, the optimum fracture half-length should be larger. Furthermore, as reservoir permeability becomes smaller, the need for fracture conductivity to create the optimum fracture is lessened.

Figs. 1.11 and 1.12 show an example economic optimization study conducted on a tight formation gas well ($k = 0.01$ md). These figures show that increasing fracture conductivity (by improving proppant type) has little economic impact, while increasing fracture half-length has significant impact. Fig. 1.12 shows that the drainage area has a significant effect on the optimum fracture half-length.

1.3.2 Fracture Mechanics. Fracture mechanics encompasses several important aspects of the fracture design process. First, fracture mechanics must consider material balance and account for the total volume of fluid injected. The material balance, therefore, must account for both the fluid that leaks off and that which remains in the fracture as the fracture is propagating. Second, the relationship between fracture width and the applied hydraulic pressure must be established. Third, the pressure loss associated with fluid flow in the fracture must be accounted for by application of fluid flow equations. Finally, the tip-propagation pressure criterion must be satisfied.

Material Balance. The material balance (continuity) equation for hydraulic fracturing simply expresses the following relationship:

Section Takeaway

- Elasticity and material balance are important to fracture design and execution.
- Fracture geomechanics can be described by the net treating pressure that is made up of several critical parameters over which we have little control (E, H, and tip effects) and those parameters that we can control but have a lesser impact on the process (Q and μ).
- 2D fracture relations can provide insight into the fracturing process; however, fracture mechanics is a complex process demanding more rigorous tools.

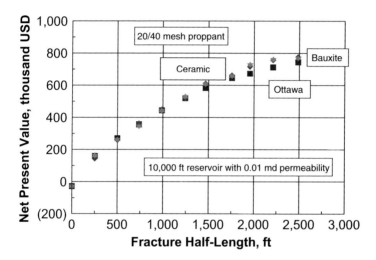

Fig. 1.11—Optimum fracture half-length as a function of fracture conductivity.

Volume Pumped = Volume in Fracture + Volume Lost. .... (1.5)

Defining volume pumped, volume lost, and volume in the fracture by V_p, V_L, and V_f, respectively, we have

$$V_p = qt_p, \quad\quad\quad\quad\quad\quad\quad\quad\quad\quad\quad\quad\quad\quad\quad\quad\quad\quad\quad\quad\quad\quad\quad\quad\quad\quad (1.6)$$

$$V_L = 2Ch_p L\sqrt{t} + 2LS_p h_p, \quad\quad\quad\quad\quad\quad\quad\quad\quad\quad\quad\quad\quad\quad\quad (1.7)$$

$$V_f = whL. \quad\quad\quad\quad\quad\quad\quad\quad\quad\quad\quad\quad\quad\quad\quad\quad\quad\quad\quad\quad\quad\quad (1.8)$$

Substituting Eqs. 1.6 through 1.8 into Eq. 1.5, and solving for the tip-to-tip length, L, gives

$$L = qt_p / (whL + 2Ch_p\sqrt{t} + 2S_p h_p), \quad\quad\quad\quad\quad\quad\quad\quad\quad (1.9)$$

where q = pump rate in ft³/min, t_p = pump time in minutes, C = fluid-loss coefficient in ft/min$^{1/2}$, h_p = permeable fracture height in feet, w = average fracture width in ft, S_p = the

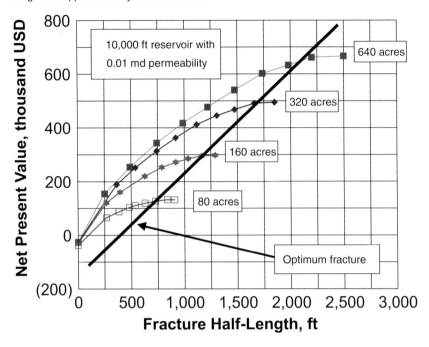

Fig. 1.12—Optimum fracture half-length as a function of drainage area.

fluid loss per area before the formation of a filter cake, and h = total fracture height in ft.

Eq. 1.9 determines the length that will result for a fracture treatment in terms of the other variables, and it approximates (within 10%) computer fracture models. Note that this equation can be rearranged to form a quadratic equation in terms of t_p. Solving this equation gives the pumping time (i.e., volume pumped) to obtain a desired fracture length. Further inspection of Eq. 1.9 indicates that increasing any of the terms in the denominator (except time) will decrease the fracture length. In particular, changing the height, h, and/or fluid-loss coefficient, C, can affect fracture length.

1.3.3 Elasticity. The width term, w, in Eq. 1.9, has caused the industry many problems because two fundamentally different assumptions that give significantly different results are used for constant height designs. The two mathematical models are commonly termed the Perkins and Kern (1961) (PK) and the Khristianovic and Zheltov (1955) (K) model. The differences in the models result from their different applications of the theory of elasticity to hydraulic fracturing. It should be noted that the PK model was later extended by Nordgren (1972), while the Khristianovic model was extended by Geertsma and de Klerk (1969). As a result, "PK" and "PKN" are used interchangeably for the Perkins and Kern model as "K" and "GDK" are for the Khristianovic model.

A classical solution in the theory of elasticity predicts that, for an infinite, elastic slab in plane strain (i.e., deformation restricted between parallel planes in the slab), with a pressurized slit through the slab, the slit will deform into the shape of an ellipse. The ellipse will have a major axis equal to the slit half-length and a minor axis proportional to the pressure and slit length and inversely proportional to the elastic modulus as seen in **Fig. 1.13a** for

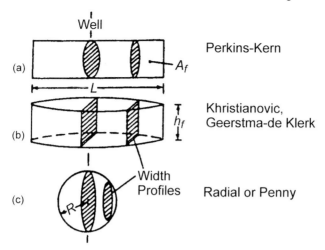

Fig. 1.13—Schematic of fracture models. From Nolte (1986b).

the PKN solution and Fig. 1.13b for the GDK solution. As shown, the ellipse in the PKN model is vertical, while the ellipse in the GDK model is horizontal. As a result, a debate has been waged during the last 30 years as to which is correct.

The prevailing thought is that the PKN model is most applicable for fractures that are long when compared to their height and that the GDK model is more applicable for fractures that are short compared to their height. In this latter scenario, a "penny frac" or a fully 3D model would be more appropriate. The "penny frac" model is shown as Fig. 1.13c. Note the elastic solutions for width and pressure can be found in the references for each model. However, for the general case, with length greater than height, the PKN model will predict less width; thus, from Eq. 1.9, the PKN model will generally predict more length. Also, the PKN model predicts that the net pressure (fluid pressure in fracture minus formation closure pressure) increases as length, L, increases, while the GDK model predicts that net pressure decreases with L. This difference in net pressure behavior is significant because bottomhole pressure measurements indicate that, if height is relatively constant and significantly smaller than fracture length, the pressure will increase as predicted by the PKN model. Also, downhole televiewer pictures obtained by Smith et al. (1982), which directly measured the fracture width in an openhole completion, indicated that the pressure-width relationship of the PKN model was most applicable. The consequence of the different width assumptions in the models is a large variation in fluid required to achieve a given fracture length with the PKN model predicting the smallest fluid requirement. As noted, the appropriate mathematical formulation for width was validated with the downhole televiewer and is continually validated today through downhole tiltmeter surveys. These have resulted in the saying, "Fractures are longer and narrower than we ever thought."

Today, fully 3D models are available, and there would appear to be no need to use pseudo-3D models. This point is illustrated in **Fig. 1.14** and demonstrates that the issue is a nontrivial one. This figure plots the ratio of 2D width calculations to fully 3D width calculations as a function of the length-to-height ratio. Fig. 1.17 highlights the differences between the 2D mathematical assumptions, thereby underlining the need to use fully 3D fracturing technology.

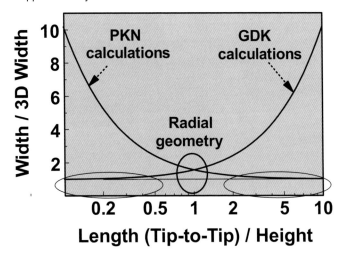

Fig. 1.14—2D or "pseudo" 3D vs. fully 3D width calculations.

1.3.4 Fracture Geomechanics. To execute a successful fracture job, the effect of several critical parameters must be understood. These parameters fall into two distinct categories: those over which we have little control but need to understand, and those that we control but that have less impact on the process. The former category includes fracture height, fluid-loss coefficient, fracture tip effects, Young's modulus, and Poisson's ratio. The latter category includes pump rate and fluid viscosity.

At this stage we note that hydrocarbon bearing and bounding formations are generally considered to be linearly elastic systems. As a result, formation properties can be characterized by two elastic constants, Young's modulus and Poisson's ratio. Young's modulus characterizes how "hard" or "stiff" the formation is and quantifies how easily a core is deformed by an axial stress. Poisson's ratio quantifies how a core expands or contracts laterally by an axial compression or tension, and in conjunction with Young's modulus characterizes the transmittal of the horizontal pressure caused by the overburden.

The importance and interaction of these categories is best understood by reviewing fracture-modeling relations. Consider, for example, the PKN model. For confined height fractures, net pressure, fracture width, and shut-in pressure-decline relations are shown below in Eqs. 1.10a through 1.10c:

$$p_{net} = (p - \sigma_c) \propto \left[\frac{E'^4}{h^4} \left(\frac{\mu q x_f}{E'} \right) + \frac{K^4_{\text{1c-app}}}{h^4} \right]^{1/4}, \quad \dots\dots\dots\dots\dots\dots \quad (1.10a)$$

$$w \propto \left[\left(\frac{\mu q x_f}{E'} \right) + \frac{K^4_{\text{1c-app}}}{E^4} \right]^{1/4}, \quad \dots\dots\dots\dots\dots\dots\dots\dots \quad (1.10b)$$

$$\Delta p^* \propto \frac{E' C h_p}{h^2}, \quad \dots\dots\dots\dots\dots\dots\dots\dots\dots\dots\dots \quad (1.10c)$$

where h is the fracture height, μ is the fracture fluid viscosity, q is the pump rate, E' is the plain strain modulus [$E/(1 - v^2)$], E is Young's modulus, x_f is the fracture half-length, $K_{\text{lc-app}}$ is the apparent fracture toughness, Δp^*, and C is the pressure decline parameter from pressure decline analysis used to determine leakoff coefficient (Gidley et al. 1989) and is related to the rate of pressure decline.

Eqs. 1.11a and 1.11b show the relations for a radial fracture, where

$$p_{\text{net}} \propto \frac{E'}{R} \left(\mu q R / E' \right)^{1/4} \quad \dots\dots\dots\dots\dots\dots\dots\dots\dots\dots\dots\dots\dots\dots \quad (1.11a)$$

$$w \propto \left(\mu q R / E' \right)^{1/4} \quad \dots\dots\dots\dots\dots\dots\dots\dots\dots\dots\dots\dots\dots\dots \quad (1.11b)$$

A review of the mathematical model relations for fracture geometry and leakoff for a confined height and radial fracture (Eqs. 1.10a through 1.10c and Eqs. 1.11a and 1.11b, respectively) shows that fracture dimensions are strongly affected by the fracture height, modulus, leakoff, and tip process, and to a lesser degree by material properties under the design engineer's control, such as pump rate and fracture fluid viscosity. It is easy to understand the importance of fracture height to the design process because the greater the fracture height, the larger the volume of fluid and proppant required to achieve the designed fracture dimensions. Control of fracture height can be more difficult, however.

Fracture height is generally controlled by the in-situ stresses and formation thickness and is difficult to measure directly. The fluid-loss variable, C, is a complex function of the fracturing and reservoir fluids, formation permeability, and fluid-loss additives and must be measured in the field. Thus, modulus becomes the only parameter that can and should be directly measured. The variable governing fracture toughness, $K_{\text{lc-app}}$, is poorly understood and difficult to measure in the field. In addition, the Poisson's ratio has little impact on the plain strain modulus because in hydrocarbon-bearing rocks, Poisson's ratio generally only varies from 0.2 to 0.3. Consequently, we may conclude that an understanding of the roles of Young's modulus and its estimate in the laboratory are important.

As shown, the net pressure is directly related to fracture height and nearly directly to Young's modulus. Also, note the limited role of viscosity, pump rate, and fracture length on net pressure. Young's modulus is also a dominant parameter in determining fracture width. Finally, Young's modulus affects the pressure decline behavior and, thus, the analysis of pressure decline data for the critical parameter, fluid-loss coefficient, C. These arguments also apply to situations where we wish to use 3D stimulations or the GDK model.

The primary factors that influence fracture height are the in-situ stress contrast, depth of invasion as a result of leakoff, the fracture pressure, and—to a lesser degree—the Young's modulus contrast between formations. Bedding, ductility, fluid gradients, and strength differences also play a minor role. The in-situ stress contrast between the target formation and bounding beds is the most important parameter, because if there is no stress contrast, there is little to nothing that can be done to control fracture height growth. On the other hand, if there is some stress contrast, the design engineer may be able to contain the fracture or limit the height growth. The ability to accomplish this objective then relies on the magnitude of the formation thickness, leakoff, and modulus related through Eq. 1.10. If the target formation is thin with little leakoff and a high modulus, then containment may be difficult even with significant stress contrast.

Section Takeaway

In-Situ Stress Contrast

In-situ stress contrast is defined as the difference in the minimum horizontal stress (fracture closure pressure) between a layer and its neighboring layer. In a typical sand-shale sequence, for example, it is the difference in stress between the sandstone and the bounding shales.

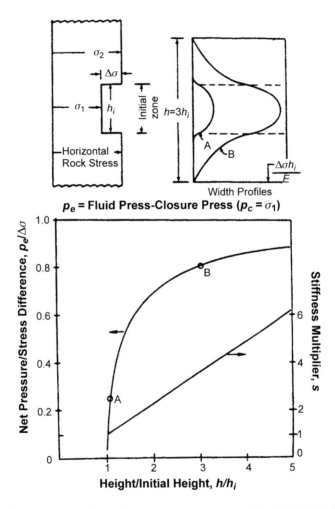

Fig. 1.15—Height-growth relationships for ideal single zone. From Nolte (1988).

The effect of closure-stress variations on fracture height growth can be seen in **Fig. 1.15.** This figure shows a plot of the ratio of net pressure to in-situ stress difference vs. the ratio of fracture height to the initial fracture height for a simple three-layer system. As shown, nearly unlimited fracture height growth occurs as the net fracture pressure reaches 85% of the in-situ stress contrast. The goal of the design engineer should be to understand the magnitude of the in-situ stress contrast and design a treatment that stays below this net-treating-pressure threshold to achieve the desired fracture dimensions. This is indeed a difficult task (see Eqs. 1.10a through 1.10c). The only things in the design engineer's control are pump rate and fracture fluid viscosity. Their effect on the net treating pressure is limited (a quarter-power effect); and reducing pump rate or fluid viscosity by half only reduces the net treating pressure by 18% if the fracture is contained, and the reduction has almost no effect in the absence of an in-situ stress contrast. A review of the published stress data in North America (more than 200 samples) indicates an average stress contrast of approximately 750 psi (0.09 psi/ft).

Finally, though Young's modulus (actually plain strain modulus, $E' = \frac{E}{1-v^2}$) has a nearly direct effect on the magnitude of the net treating pressure (as shown in Eqs. 1.10a through 1.10c), the effect of modulus contrast on fracture height growth between formation and bounding beds is usually small. For example, calculations suggest that a ten-fold difference in modulus results in a tip-to-tip length that is only 1.7 times the fracture height.

In the preceding discussion, we have used 2D models to understand the process and parameter relationships. But as noted earlier, the use of fully 3D models to design fracture treatments is recommended.

1.3.5 Fracturing Materials. The primary objective of hydraulic fracturing is to create a conductive pathway through which hydrocarbons can be produced to the wellbore. Achievement of this objective requires the use of two principal materials: fracturing fluid (and additives) and proppant. The purpose of the fluid system is to initiate, propagate, and transport the proppant, while the purpose of the proppant is to maintain the conductive pathway to the wellbore once the pumps are shut down and the well is returned to production. The volumes and amounts of these two major materials must be specified as part of a fracture design.

1.3.6 Fracturing Fluid Systems. The selection of a fracturing fluid is subjective and involves personal experience, geographic considerations, and advice from service providers. In addition, the design engineer should consider the more specific factors geared to treatment objectives, formation, and fluid properties. The selection criteria can be grouped into the following categories:

1. Safety and environmental compatibility
2. Compatibility with formation and additives
3. Simple preparation and quality control
4. Low pumping pressure
5. Appropriate viscosity
6. Low fluid loss
7. Flowback and cleanup (for high fracture conductivity)
8. Economics

Safety and Environmental Compatibility. Safety and environmental compatibility are primary considerations in the selection of fracturing fluids. As a result, hydrocarbon-based fracturing fluids have seen dramatically reduced usage in the United States in recent years. Special precautions must be used (Ely 1985; *RP 39* 1983). Foamed fluids, either hydrocarbon or water-based, can be dangerous.

Section Takeaway

- The fracturing fluid should be selected on the basis of design and economic objectives of each application.
- Keep the fluid system as simple as possible for best treatment execution.
- No one fluid is best for all fracturing applications.
- Fluid and formation compatibility tests should be conducted for each fracturing application area. At a minimum, clay control, KCl concentration, and pH should be tested.
- Establish and communicate quality control procedures to stakeholders.
- Conduct pilot tests of the fracturing fluid in the laboratory.
- Conduct additive sensitivity tests in the laboratory.
- Validate pilot tests in the field with source water and treatment chemicals.

There are varying degrees of toxicity associated with fracturing fluid components such as methanol, frac-oil, biocides, surfactants, and crosslinkers. A breathing apparatus is required for blender operators to prevent exposure that can do irreversible brain damage. Oxidizers, such as ammonium and sodium persulfate, cannot be allowed to contact fuel sources. Corrosive acidic and basic additives should be handled with care, and material safety data sheets (MSDS) should be made available on location by the service provider and reviewed for all chemicals on location. Safety meetings should be held and all hazards discussed prior to the fracture treatment.

Compatibility With Formation and Chemical Additives. A primary consideration in the selection of a fracturing fluid system is its compatibility with the formation, formation fluids, and the chemical additives. A fracturing fluid is a complex mixture of chemicals, and a number of reactions may result with the formation fluids and the formation: swelling or migration of clays, dissolution of cementing material, ion exchange, wettability alteration, pH alteration, production of hydrogen sulfide if temperatures are low (below 180°F), production of sludges, formation of emulsions, adsorption, or plugging. Each of these factors must be evaluated, and adjustments must be made.

Simple Preparation and Quality Control. The fracturing fluid composition should be as simple as possible, since every component and additive adds to the cost and increases the burden of monitoring chemical quality, chemical addition, and mixing. Viscosifiers improve proppant transport. Buffers are used to adjust the pH for hydration, crosslinking, and thermal stability. NaCl, KCl, or cationic polymers prevent clay swelling and clay migration. Liquid hydrocarbon fluid-loss agents reduce water loss to the formation. Breakers enhance polymer degradation. Surfactants are used for better load recovery and as an aid in preventing oil/water emulsions in the reservoir. Biocides prevent biodegradation of the fracturing gel and contamination of the well. More treatments are pumped with liquid additives, especially offshore, because these additives are easier to handle and measure on the fly while pumping the treatment.

Low Pumping Pressure. Most fracturing fluids have the desirable property of being drag-reducing (having a lower friction pressure) when pumped under turbulent conditions. High-molecular-weight polymers and hydrocarbon-soluble polymers are used in water- and oil-based fracturing fluids to reduce friction pressure. The effects of poor friction-reduction capabilities in a fracturing fluid are higher treating pressure, higher horsepower requirements, and higher cost.

Appropriate Viscosity. Fracturing fluids are formulated to initiate and propagate a fracture, create sufficient width, transport proppant, and create a conductive proppant pack. Excessive fluid viscosity can increase the fracturing pressure, costs, and fracture height, while negatively impacting fracture conductivity.

Fluid viscosities should be sufficient for adequate proppant transport; however, proppant has been pumped with very-low-viscosity fluids, including water, with friction reducer.

When using thin fluids to transport proppant, such as slickwater or uncrosslinked polymer solutions at elevated temperatures, it is probable that a settled bank will form along the bottom of the fracture. Research by Biot and Medlin (1985) and Medlin et al. (1985) indicates that the formation of an equilibrium bank may not apply to field fractures, although such banks have been observed in laboratory-scale slot-flow devices.

Fluid-Loss Control. Fracturing fluid systems offer varying degrees of fluid-loss control. Water-based fluids with polymer give fluid-loss control by building filter cake as the fluid leaks off into formations with permeability less than 5–10 md. Foams with gas-internal phases can give fluid-loss control comparable to gels with hydrocarbon when the liquid external phase of the foam is stabilized with polymer. Leakoff with foams is also dependent on formation permeabilities.

At reservoir permeability of less than 0.1 md, the total fluid-loss coefficient, C_t, starts to become influenced by leakoff resistance in the reservoir rock. Reservoir-leakoff resistance is influenced by the fracturing fluid filtrate and the formation fluids, the porosity of the reservoir, the compressibility of the formation fluids, and the leakoff-driving pressure (the fracturing pressure minus the reservoir pressure), as well as the reservoir permeability. Increasing the leakoff driving pressure from 500 to 2000 psi can increase C_t by a factor of 3. For a reservoir permeability of less than 0.01 md, the reservoir resistance dominates leakoff, and the fluid leakoff properties are no longer important. Little added fluid-loss reduction is gained by using fluid-loss additives.

For naturally fissured reservoirs, it is of primary importance not to allow polymer to leak off into and plug the natural fractures if the fissures provide the primary source of reservoir permeability. Fluid-loss additives—particulate agents such as 100-mesh sand—that prevent polymer from entering the natural fissures should be considered.

The spurt loss of a fluid, which is defined as the fluid loss per area before the formation of a filter cake, can be significant in naturally fractured reservoirs as well as in reservoirs with permeability in the range of 1 to 10 md. Spurt loss increases strongly with reservoir permeability and leakoff driving pressure and is affected by factors that affect fluid flow in reservoirs, such as filtrate viscosities and compressibility.

Flowback and Cleanup. To achieve a conductive fracture, the fracturing fluid must be removed from the formation. As discussed above, it is essential to prevent polymer from invading the rock matrix and natural fissures. Good fluid-loss control and fluid breaker can accomplish this.

Gel breakers oxidize the polymer backbone, enabling the polymer to be produced out of the fracture. Ammonium and sodium persulfate are commonly used at temperatures above 150°F, or at lower temperatures with an activator. Enzyme breakers such as hemicellulose are used at temperatures below 120°F with a pH less than 8.5. In 1991, service companies began offering encapsulated or crushable breakers designed to release the oxidizer after pumping has stopped.

Encapsulated breakers are used very aggressively in the pad stage (e.g., 7 lbm/1,000 gal) (Small et al. 1991), with the concentration reduced somewhat during later proppant stages.

Sometimes a conventional breaker is added at the final stages of the treatment to enhance near-wellbore cleanup. The benefit of adding large amounts (greater than 2 lbm/1,000 gal) of conventional breaker is suspect. Tests have shown that a frac fluid with 2 lbm/1,000 gal breaker can be broken prematurely before it gets down to the fracture.

Fewer problems with fracture conductivity impairment result when using hydrocarbon-based fluids or foams, as long as they break properly. Hydrocarbon gels are broken with base additive. At lower temperatures (e.g., <120°F), breaking hydrocarbon gels can be a problem. Foams break when the liquid drains, when the surfactant adsorbs onto the rock, and/or when the polymer in the liquid phase breaks. Flowback with foams has the added advantage of the nitrogen or CO_2 expansion.

Economics. After narrowing the list of possible fracturing fluid systems, the engineer should compare their relative costs. The costs of the base fluid and additives should be tallied along with disposal costs. For hydrocarbon-based fluids and polymer emulsion, the value of recovered hydrocarbon should be considered. Hydrocarbon fluid and foam treatments are considerably more expensive because of the added safety, equipment, and implementation costs.

In addition to the costs of materials and pumping, one should consider the net present value of post-frac production. This is a function of the fracture geometry and conductivity. This evaluation is best done using an integrated design package including a fracturing simulator, a production simulator, and an economic optimization program.

1.3.7 Proppants. Proppant is used to prop open the fracture and provide a conductive pathway from the reservoir to the wellbore. Its effectiveness is evaluated by the magnitude of the achieved fracture conductivity.

Types of Proppants Available. There are several classes of proppants currently available. The most common is sand. Resin-coated proppants are an alternative. These proppants begin with either sand (the most common) or ceramic proppants, and a thin coating of resin is added. These proppants are differentiated primarily by their specific gravity and strength. The cost of ceramic proppant ranges from 5 to 10 times that of sand.

Proppant Size. Proppant particle size is a critical design consideration that depends on the stress level, desired conductivity, and proppant transport (i.e., achievable fracture width). In general, larger proppant is better, because conductivity is increased with minimal impact on the cost if it can be placed successfully in the desired fracture treatment.

Section Takeaway

- Proppant type, size, and concentration are all important to generating fracture conductivity.
- TSO fracturing can improve fracture conductivity, provided that the TSO occurs while the proppant slurry is mobile and unit slope behavior on a Nolte-Smith plot is exhibited.
- Proppant stress (Eq. 1.15) should be considered in fracture design.
- The maximum stress on the proppant occurs early in the well life (during cleanup if flowback control is not maintained) near the wellbore.
- Fracture fluid cleanup is critical to achieving effective fracture dimensions.
- Multiphase flow can dramatically reduce the effective fracture conductivity.

When proppant bridging caused by small perforation diameter, narrow fracture width, poor proppant transport, formation embedment, and/or fines invasion occurs, larger proppant does not necessarily improve the fracture conductivity. In this latter case, a smaller proppant diameter may give better "formation control." In most cases, 20/40, 16/20, or 12/20 mesh proppant is used, though smaller 30/60 or 40/70 mesh is often used in "frac-pack" completions in unconsolidated formations where formation sand control becomes an additional primary consideration for proppant selection. **Fig. 1.16** shows a plot of conductivity as a function of closure stress for various proppant types with similar mean particle diameters. As shown, for up to 2,000 psi of closure stress, the conductivity of the proppants is similar. However, the conductivity of the various proppant types reacts differently as closure stress is applied.

The size distribution can play a significant role in the fracture conductivity, with a narrow size distribution generally providing improved conductivity over a larger distribution with more fines. Testing of proppant size distribution requires a sieve analysis, which should be performed periodically as a quality-control practice on fracture treatments. The American Petroleum Institute (API) provides two publications detailing tests for sands and intermediate- and high-strength proppant (*RP 56* 1983; *RP 60* 1989). API recommends that a minimum of 90% of the tested sand sample fall between the designated sieve sizes correlative to the indicated mesh size (e.g., 6/12, 12/20, 20/40). Less than 0.1% of the total proppant sample should be larger than the largest sieve screen mesh, and no more than 1.0% should be smaller than the smallest sieve screen mesh (i.e., less than 1% fines). Fines of any form from any source are a major concern for proppant. The presence of silts, clays, and any other fine particles in the proppant will reduce fracture conductivity.

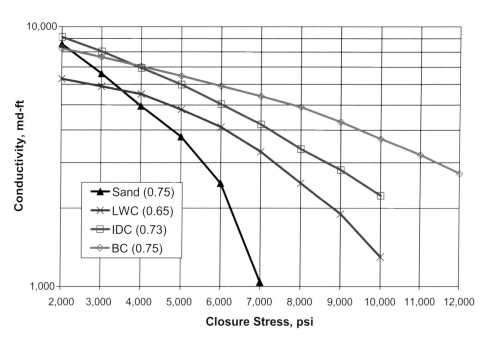

Fig. 1.16—Example of baseline conductivity data. From Schubarth (2004).

Proppant Concentration. Another parameter for consideration in generating fracture conductivity is proppant concentration. The hydraulic width of a fracture (i.e., PKN) is proportional to the product of net treating pressure and fracture height divided by Young's modulus. Fracture width also is proportional to the ratio of the fourth root of the product of pump rate, viscosity, and fracture half-length. As a result of the quarter-power effect, there is little to do to alter the hydraulic fracture width. The final propped width is then related to the hydraulic width and the final in-situ proppant concentration. For example, according to API *RP* 56, if the hydraulic width is 0.2 in. and sand is used as proppant with a final in-situ concentration of 9 lbm/gal, the "proppant coverage" is 0.8 lbm/ft^2. This shows that even achieving 1 lbm/ft^2 is difficult, but the standard test data is collected at 2 lbm/ft^2. Thus, typical in-situ fracture conductivity is generally less than one-half of the measured laboratory data.

Several additional parameters are important to the generation of fracture conductivity with proppant. Sphericity and roundness are two such parameters. Sphericity refers to how well a proppant grain is approximated by a sphere, while the roundness is a qualitative measure of the proppant surface smoothness. The most widely used measure of sphericity and roundness was developed by Krumbein and Sloss (1963). API recommends that sand used as a proppant should have both roundness and sphericity of 0.6, while a ceramic proppant should have a minimum value of 0.7. For most applications with proppant stress in excess of 4,000 psi, the more spherical the proppant, the higher the permeability. At lower stresses, the angularity of the proppant actually improves the permeability associated with nonuniform packing. These poorer-quality angular proppants should not be used in higher-stress applications because they typically have more associated-fines-generating materials and tend to crush more easily when they come in contact with near-wellbore stresses.

TSO Fracturing Technique. Another means of generating fracture conductivity is the TSO fracturing technique, which is used to halt fracture propagation (height and length) while still pumping a mobile proppant slurry to generate additional fracture width, thereby increasing fracture conductivity (Smith et al. 1984; Martins 1992). To execute such a design, the low-proppant concentration stage is extended over a conventional treatment design so that it reaches the tip of the fracture as the pad depletes, fracture propagation ceases, and additional slurry is pumped increasing the fracture width. **Fig. 1.17** shows a TSO design schematic highlighting the extended low concentration and mobile slurry stages.

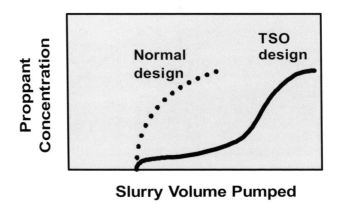

Fig. 1.17—TSO design schematic.

TSO fracture designs can effectively increase fracture conductivity by increasing fracture width. Because the design is predicated on a fracture storage phenomenon, width increases associated with a TSO design are readily determined as a function of fracture volume or time. **Fig. 1.18** illustrates such a phenomenon (Nolte 1986). As shown, a two-fold increase in fracture width and conductivity can be achieved for a 50% efficiency treatment with a 1.8-fold increase in the treatment size.

The relationship between fracture volume and width (defined by Fig. 1.22) is but one of the critical TSO design criteria. Additional parameters of importance to TSO designs include the fracture geometry, length, and Young's modulus. *Fracture stiffness* is defined as the rate of change of the mean net pressure with fracture volume at a constant fracture area. Simply stated, fracture stiffness is the formation's resistance to the width growth that occurs during a TSO design. Fracture stiffness is

$$s = \left(\frac{dp_{net}}{dV_f} \right) A. \dots\dots\dots\dots\dots\dots\dots\dots\dots\dots\dots\dots \quad (1.12)$$

For confined height, fracture stiffness becomes

$$s = \left(\frac{E'}{\pi h_f^2 L_f} \right), \dots\dots\dots\dots\dots\dots\dots\dots\dots\dots\dots \quad (1.13)$$

while for radial growth fractures, stiffness becomes

$$s = \left(\frac{3E'}{16r^3} \right) \dots\dots\dots\dots\dots\dots\dots\dots\dots\dots\dots\dots \quad (1.14)$$

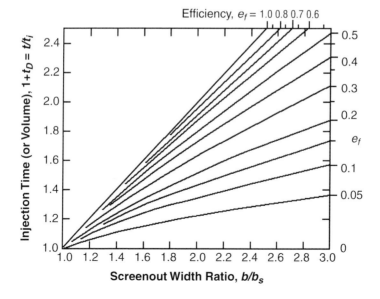

Fig. 1.18—TSO width as a function of slurry volume and efficiency.

Inspection of Eqs. 1.13 and 1.14 show that the stiffness and, thus, the pressure rise after screenout is greatest in high-modulus formations with short fractures. This is not to imply that TSO fracture designs should not be used in these environments, but rather that consideration should be given to the rapid pressure rise that will occur after screenout and to any other implications on operations. Even in the event of a fully locked-up screenout, the production benefits realized can more than offset operational costs incurred as a result of the screenout, provided that the pressure is driven up by mobile slurry (as discussed previously) and identified with a prolonged unit slope behavior on a Nolte-Smith plot.

The next few parameters are also significantly important to the achievement of adequate fracture conductivity:

1. Proppant stress
2. Effective conductivity (fracture fluid cleanliness)
3. Proppant embedment
4. Multiphase flow
5. Non-Darcy flow considerations

Proppant Stress. The single most important factor when selecting a proppant is the stress that the proppant will experience. This stress can be readily calculated with a modification to Eaton's equation to include flowing bottomhole pressure and final net pressure:

$$\text{Proppant Stress} = [v/(1-v)](\sigma_{ob} - p) + p + p_{tec} + p_{fnp} - p_{fbhp}, \quad \dots\dots\dots\dots \quad (1.15)$$

where v = Poisson's ratio, σ_{ob} = overburden stress (approximately 1 psi/ft of depth), p = reservoir pressure, p_{tec} = stress caused by tectonics (usually unknown and omitted), p_{fnp} = final net treating pressure, and p_{fbhp} = flowing bottomhole pressure, which is the post-fracture ISIP minus the minimum horizontal stress. For a sandstone reservoir ($v = 0.25$) at 10,000 ft of depth, normal pressure gradient, final net treating pressure of 1,000 psi and flowing bottomhole pressure of 500 psi, an estimate of proppant stress is 6,722 psi (if we assume the overburden is 1 psi/ft and neglect the tectonic component).

A few observations can be made by studying this equation. First, as the reservoir pressure is depleted, the stress on the proppant decreases. Second, as the well is drawn down further (p_{fbhp} becomes smaller at the wellbore), the stress on the proppant will increase. Also, because pressure in the fracture is higher away from the wellbore, the maximum stress that a proppant will see is early in the life of a well near the wellbore. In fact, in gas wells, the maximum stress on the proppant is most likely to occur during well cleanup when gas breaks through, unless flowback operations are closely monitored.

Laboratory data are available for every commercial proppant as a function of proppant stress. In addition, crush-resistance tests can be performed on any proppant for quality assurance.

Effective Conductivity (Fracture Fluid Cleanliness). Inefficient and ineffective fracture fluid cleanup is one of the most important aspects of a fracturing treatment. For many years, the relationship between load recovery and post-fracture productivity has been known (Soliman and Hunt 1985), but only recently have the various mechanisms affecting cleanup been characterized.

Research has identified six major contributing mechanisms to the reduction of proppant pack permeability. These mechanisms are

- The effects of time and temperature on proppants (Samuelson and Constein 1996)
- Gel (solid) residue in the proppant pack that leads to damage (Penny and Jin 1996a, 1996b; Shah et al. 1992)
- Viscous fingering through the proppant pack (Pope et al. 1994)
- The existence of unbroken fluid on proppant pack permeability (Voneiff et al. 1996)
- Non-Darcy and multiphase fluid flow effects (Penny and Jin 1996c; Holditch and Morse 1976)
- Capillary pressure (water block) effects (Holditch 1979)

A large number of studies have looked at the cleanup of a fracture following a treatment. In situations where crosslinked, non-Newtonian fracturing fluids are used, evidence suggests that fluid leakoff results in significant changes in the rheological characteristics of the injected polymer. Specifically, a threshold pressure gradient needs to be overcome before the polymer in the fracture will flow. Cleanup that occurs is a function of the threshold gradient in the post-leakoff gel. Small volumes of residue that are left behind affect the porosity and permeability of the pack significantly (Pope 1994). Field observations also indicate that the level of cleanup is a function of the amount of free water produced and that the onset of gas production slows cleanup because of multiphase effects; thus, the delay in gas production helps in cleanup (Pope et al. 1996; Anderson et al. 1996; Willberg et al. 1997).

Taken together, studies show that the creation of fractures with dimensionless fracture conductivities in excess of 10 fully clean up over the entire created length (May et al. 1997; Neghaban et al. 1997; Ayoub et al. 2006).

Proppant Embedment. The term *embedment* describes the lost conductivity caused by the interaction between the proppant and the formation at the face of the fracture. Although primarily a problem in lightly consolidated to unconsolidated formations, it can and does occur in hard rock as well to a lesser degree. **Fig. 1.19** shows a scanning-electron-microscope view of the proppant/formation interface under confining pressure. As shown, nearly a third of a grain of diameter is lost as a result of embedment into the formation. A loss in conductivity because of embedment is usually not a significant matter.

Multiphase Flow. Multiphase flow (Penny and Jin 1995; Jin and Penny 2000; Milton-Tayler 1993) can dramatically and detrimentally impact the effective conductivity of a fracture. This flow effect has been described with relative permeability curves as shown in **Fig. 1.20.** The curve shows dramatic reductions in gas flow with even small amounts of water flowing in the fracture. This effect is even more dramatic in the presence of three flowing phases (oil, gas, and water). As a result, multiphase flow can be a major detriment to post-fracture well response.

Non-Darcy Flow Considerations. Flow down a fracture may not be adequately described by Darcy's law if the flow velocity is high. Non-Darcy flow reduces the effective permeability of the proppant pack to the flowing fluid (Cooke 1973; Guppy et al. 1982a, 1982b; Settari et al. 2002; Stark 1998; Holditch and Morse 1976; Schulte 1986; Bale et al. 1994). Smith et al. (2004) present several examples that demonstrate the influence of non-Darcy flow.

1.4 Prefrac Data Collection

1.4.1 Formation Evaluation. Important parameters to the characterization of a reservoir for hydraulic fracturing and hydrocarbon recovery include the porosity, permeability, fluid saturations, and lithology of the formations. Knowledge of these parameters is used to

100 UM ▬

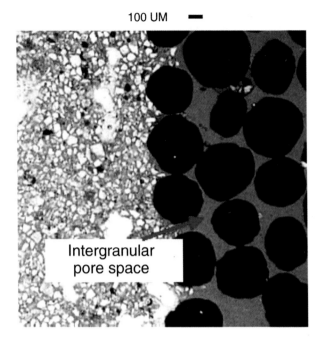

Intergranular
pore space

Fig. 1.19—SEM photoplate of proppant-formation interface.

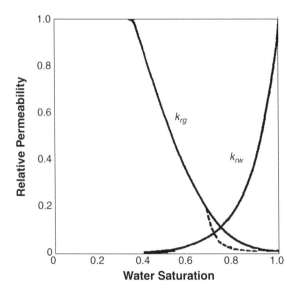

Fig. 1.20—Fracture relative permeability curves (after Holditch 1979).

estimate net pay, in-place hydrocarbons, and pre- and post-fracture rate of recovery. A subsequent section on logs, cores, and pressures summarizes their use in fracture stimulation.

1.4.2 Logs. The dipole sonic log (any long-spaced log with shear and compressional travel times) has been shown to have application in the estimation of the minimum horizontal stress distribution (Dutton et al. 1982). These data are critical in defining the differential closure stresses between beds for determining fracture height growth parameters. The shear and compressional sonic velocities may be used to calculate dynamic elastic rock properties and theoretical closure stress in a given horizon. Stresses calculated with this log should be calibrated to actual in-situ stresses by measuring the closure stress in three or four zones in the wellbore and then realigning the calculated stresses in both sandstones and shales.

Several methods currently exist to determine fracture azimuth. The most commonly used techniques are borehole breakout analysis, post-frac shielded gamma ray logging, surface tiltmeters, and microseismic mapping. Strain-relaxation techniques may also be used. Though effective, these techniques require an oriented core and specialized equipment at the wellsite.

Borehole geometry logs measure hole eccentricity or ellipticity and its orientation, and therefore must be run in an open hole. It has been noted in some fields that wellbore breakouts create elliptical cross sections, with the long axis of these noncircular sections

---

Section Takeaway

- Logs, cores, and transient pressure analysis should be utilized to establish prefrac reservoir characteristics.
- Dipole sonic logs can aid in in-situ stress estimation but must be calibrated with actual closure stress information.
- Borehole breakouts and radioactive tracers with shielded gamma ray tools can be used to determine fracture orientation (maximum horizontal stress).
- Temperature and radioactive tracer logs can be used to "infer" fracture height since they only "see" near the wellbore. Thus, they are best used when the fracture follows the wellbore (i.e., $\sigma_{hmax} >>> \sigma_{hmin}$ in essentially vertical wells).
- Triaxial compression tests should be conducted on core samples (whole core or rotary sidewall core plugs) to determine Young's modulus.
- Dynamic elastic properties should be corrected to static values for use in fracture design and analysis.
- Tiltmeter analysis assumes elastic deformation of the Earth through opening of a hydraulic fracture.
- Surface tiltmeter mapping can identify fracture orientation (vertical or horizontal) and fracture direction.
- Downhole tiltmeter mapping can identify fracture height.
- Microseismic mapping measures the shear slippages that occur due to increased stress caused by increasing pore pressure and formation stress at the fracture tip.
- Microseismic mapping can determine fracture direction, fracture length, and fracture height.

sharing a common azimuth. In cases where the minimum hole diameter is equal to the bit diameter, such washouts or spalls have been termed "breakouts" (Gough and Bell 1981; Thorpe and Springer 1982; Babcock 1978). "Breakouts" should not be confused with common washouts. The geometry of the borehole (ellipticity) may be affected by the stresses in the Earth in the near-wellbore region. The fracture azimuth is also affected by these stresses (Brown et al. 1980; Bell and Gough 1979) as well as the shear strength of the rock (Bell and Gough 1979). Unequal stresses will cause a preferential stress concentration on the side of the wellbore perpendicular to the maximum stress direction, and if the shear strength is high enough, breakout will be limited to this region. In such a case, the breakout will develop with the long axis of the elliptical borehole perpendicular to the expected azimuth of hydraulic fractures.

To determine azimuth by a shielded gamma ray log with a gyroscope, the fracture treatment is traced with radioactive tracer. The fracture direction is determined by identifying the direction of the maximum gamma ray count. This method works best in openhole environments because the results are strongly influenced by perforation phasing.

Temperature logging is often used to infer fracture height. To be effective, a base temperature log must be run to determine geothermal gradient and static bottomhole temperature. To obtain a valid static temperature survey, the well should be shut in for several days prior to logging. Temperature disturbances caused by circulating the well during cleanout operations, etc., usually require several days to dissipate.

Temperature decay profile surveys should be run as soon as possible after a fracture injection.

A minimum of three logging runs should be made at intervals of 45 minutes from the start of each run. No backflow from the well should be allowed prior to or during temperature profiling. The logs should be run from several hundred feet above the pay interval to several hundred feet below the fracture bottom or to plugback total depth.

After a minifrac or fracture treatment, heat transfer will occur above the treated zone by radial heat conduction, while over the fracture faces, heat transfer will be by linear flow. Ideally, across these two areas, temperatures will recover at different rates following the end of pumping, causing a temperature anomaly to develop that identifies the fractured zone. Unfortunately, this ideal situation rarely occurs, making misinterpretation of post-frac temperature logs all too common.

A static base temperature log and cold-water circulation survey may be run to determine the temperature gradient and identify anomalies caused by formation changes, the wellbore, and the completion. **Fig. 1.21** shows the conductivity effects from different formations on both pre- and post-frac logs (Dobkins 1981). Note that a washout behind casing will create a cool anomaly that may be interpreted as a fractured zone. On the other hand, a washout completely filled with cement will insulate the wellbore and create a "hot nose" on the log. Also, a change in tubular diameter such as the bottom of tubing or casing can cause an "offset" in the log. All of the above anomalies can be detected with the base temperature log and removed from the post-frac log interpretation.

Fig. 1.22 shows a warm anomaly or hot nose above the fractured zone and the obvious problems associated with picking the fracture top (Dobkins 1981). It has been theorized that this is caused by fluid movement after shut-in and that the hot nose is part of the fracture height.

Temperature crossovers are often seen below the perforated interval from one logging run to another. Below the perforations, the wellbore is filled with stagnant, hot fluid. Any downward

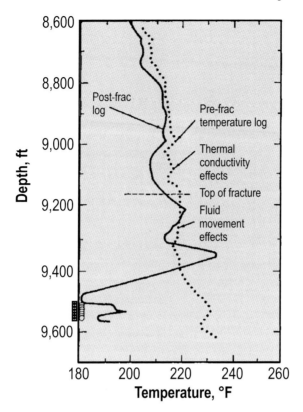

Fig. 1.21—Pre- and post-fracture temperature logs showing thermal conductivity effects.

fracture growth will place fluid outside the casing that is cooler than the fluid inside. Thus, heat flow will be in the direction opposite that from across and above the fractured zone, and the wellbore may cool down with time. This often results in a temperature "crossover," as seen in **Fig. 1.23,** which can be a good indicator of the bottom of the created fracture.

Temperature logs are shallow investigative tools; thus, they work best if the created fracture is vertical. For the same reason, this log is a poor tool to run if the fracture is vertical and the wellbore is deviated. This is especially true if the horizontal stresses are nearly equal (hoop stresses around the wellbore are large), driving the fracture away from the wellbore and making fracture geometry difficult to interpret from a temperature log.

If the well were to "go on vacuum" after a stimulation, the falling fluid level would continually carry warm fluid downward into the fractured zone, obscuring the temperature anomaly. This is possible in injection well stimulations and on pumping wells with low reservoir pressure. In such cases, the fluid level should be allowed to stabilize before running the logs.

In addition to temperature logging, post-frac gamma ray logs are often run to evaluate fracture height. To conduct this evaluation, the mixture of fracturing fluid and proppant is tagged with radioactive materials with varying half-lives. Noting the variation in half-lives, a post-frac gamma ray log should be run early in the half-life of the tracer used. Also, for the most definitive results with regard to fracture height, the tagged material should be added throughout the stimulation.

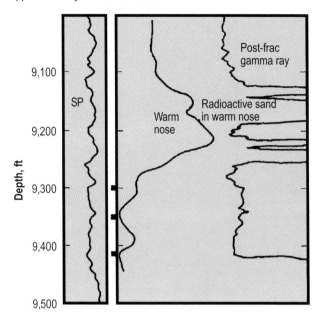

Fig. 1.22—Temperature logs showing warm anomaly above treatment zone.

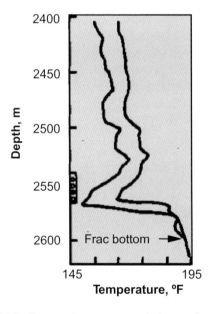

Fig. 1.23—Temperature crossover below perforations.

One advantage of gamma ray logs over temperature logs is that they do not need to be run immediately after stimulation. However, the other restrictions on the temperature logs apply equally to radioactivity logs. Thus, while the two logs are often used in combination, the potential exists for them to confirm one another and still not yield reliable results.

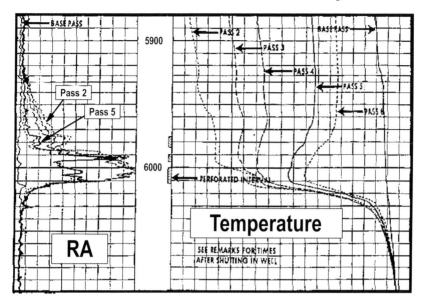

Fig. 1.24—Post-frac temperature and gamma ray log.

Radioactivity logs cannot distinguish between a fracture and a small channel behind the casing, because any material deposited in a channel is indistinguishable from tagged material in a fracture.

Fig. 1.24 shows an example of pre- and post-frac gamma ray and temperature logs (Dobkins 1981). In this example, the radioactive material was used in only the later part of the fracture treatment; thus, radioactive material appeared only through a portion of the fracture (Brown et al. 1980). In this same figure, radioactive material appears across the hot nose, indicating this to be, in fact, part of the fracture height. Today, multiple radioactive isotopes (tracers) are available for use throughout the fracture stimulation, allowing the design engineer to assess through which perforations the various fluids and proppant stages entered the fracture.

1.4.3 Core. Core analysis is the best technique available for obtaining Young's modulus and Poisson's ratio for use in design and post-appraisal. In fact, the modulus of the formation is the only fracturing parameter that can be directly measured through triaxial compression tests. This measurement improves the analysis of the other geomechanical parameters (in-situ stress contrast and leakoff coefficient).

Both elastic constants should be measured in the laboratory with a triaxial compression test. This test characterizes the elastic properties under static conditions (a slowly varying load) similar to that resulting from the hydraulic fracturing process. **Fig. 1.25** shows a plot of effective axial stress vs. strain resulting from a triaxial compression test. This test is generally conducted by subjecting a small jacketed core plug to confining pressure and loading the sample axially to produce plots of axial, lateral, and volumetric strain in excess of confining pressure (effective axial stress). Shown on this figure is a line drawn through the linear elastic region, the slope of which is Young's modulus. Also shown is the deviation from this linear elastic region, generally at pressures in excess of those seen during

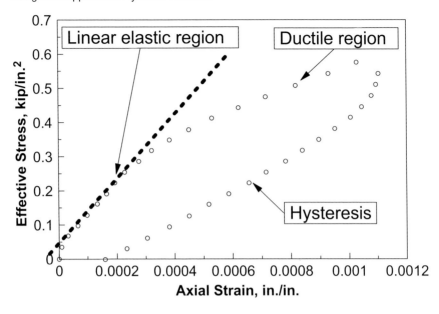

Fig. 1.25—Stress strain data; from Smith et al. (2001).

fracturing, which represents ductile failure (grain rearrangement) of the sample and, finally, sample failure. This core sample represents a (soft sediment) lightly consolidated to unconsolidated formation ($E = 1 \times 10^6$ psi). In a harder, more consolidated formation, the size of the ductile region would be much smaller and might not even be evident on the plot because deviation from linear elastic behavior and sample failure is small.

In contrast, dynamic values of the elastic properties can be inferred (using elastic relations) from the compressional and shear travel times from sonic logs. The difference between the dynamic and static modulus can be significant and can require correlations to relate the dynamic properties to the static properties needed for design. **Fig. 1.26** shows an estimate of dynamic modulus derived from compressional travel times, while **Fig. 1.27** shows a correlation of dynamic Young's modulus to static Young's modulus for the south Texas Wilcox formation (Britt et al. 2006).

The use of oriented cores to predict fracture azimuth has been suggested for many years (Brown et al. 1980; Bell and Gough 1979). The chief advantage of core analysis for fracture azimuth is its ease of application. During routine coring operations, the additional work required to orient and analyze the core is small compared to that of other azimuth-measuring procedures (Rowley et al. 1981; Robertson 1964). Also, since most coring is done early in the life of the field, the azimuth data collection is very timely. The biggest disadvantage to common oriented core analysis is the fact that this is an indirect measurement, and it is difficult to be certain that the answer is correct. The most successful core analysis is the direct on-site measurement of strain relaxation (Blanton 1983; Blanton and Teufel 1983).

The strain-relaxation technique has proved to be accurate in several tests in which azimuth was also measured with other procedures (Lacy 1987; Teufel 1982; Teufel et al. 1984; Smith et al. 1985). These include tests in a volcanic tuff in Nevada; a low-permeability Mesaverde sandstone; a low-permeability gas sand in the Cotton Valley formation; and a high-porosity, high-permeability sandstone in Oklahoma.

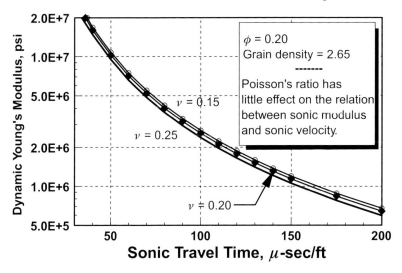

Fig. 1.26—Dynamic modulus from compressional travel time.

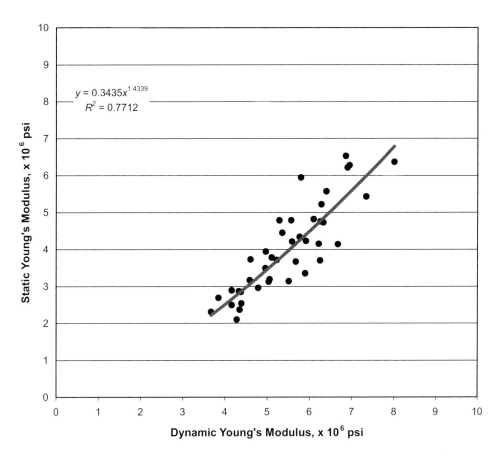

Fig. 1.27—Dynamic to static modulus correlation (south Texas, Wilcox formation).

1.4.4 Surface Tiltmeters. Tiltmeters are highly sophisticated, extremely accurate biaxial instruments that use "bubble" sensors to measure the change in angle of a surface. The use of tiltmeters to monitor hydraulic fractures, at depths as great as 12,000 ft, is based on the assumption that the Earth will respond in an elastic manner to deformations caused by opening a hydraulic fracture. In that case, the surface of the Earth will deform in a predictable manner, and measurements of this deformation can be interpreted to obtain data with respect to fracture geometry (Wood et al. 1976; Wood 1981; Davis 1983). **Figs. 1.28 and 1.29** illustrate surface deformations associated with vertical and horizontal fractures, respectively.

A typical tiltmeter array consists of 12 to 16 instruments evenly spaced radially around the well, at a distance of about 0.4 times the depth of the zone to be fractured. Each instrument is installed in a shallow-cased hole, usually 10 to 20 ft deep, and packed into position using sand to insulate the device from surface weather and noise effects. **Fig. 1.30** shows a surface tiltmeter site (PVC pipe) at a western Missouri project (Britt et al. 2006).

The tiltmeter instruments are capable of measuring changes in tilt of a surface with accuracy of about 1×10^{-7} radians. Because of the sensitivity of the measurements, changes

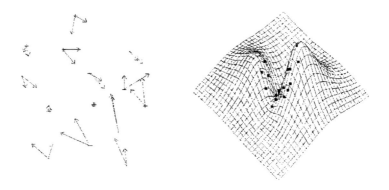

Fig. 1.28—Surface tiltmeter monitoring (vertical fracture).

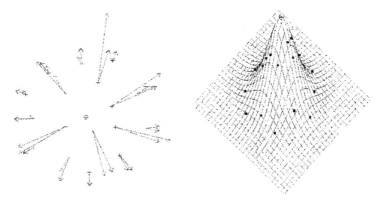

Fig. 1.29—Surface tiltmeter monitoring (horizontal fracture).

Fig. 1.30—Surface tiltmeter location in western Missouri.

in the level of the Earth's crust associated with solid-Earth tides cause changes in the surface angle that are orders of magnitude greater than the fracture treatment. Fortunately, the period of the fracture event is much shorter than the "tidal noise" and can be separated by post-analysis using frequency domain filtering and/or tidal filtering. The residual from this filtering is then used to measure the tilt signal related to hydraulic fracturing. The signals from both channels of a tiltmeter are combined to form a tilt vector that embodies direction and magnitude of the tilt measured at that site.

To analyze the data, observed tilts are compared with theoretical values for many possible combinations of fracture azimuth and dip; thus, the azimuth and dip are determined to see which produces the least error. **Fig. 1.31** shows theoretical tilt responses for vertical, inclined, and horizontal fractures.

Just as the pattern, or direction of the tilt vectors, is related primarily to the fracture azimuth and dip, the magnitude of the vectors is principally a function of fracture volume.

Because extensive site preparation is required to install the tiltmeter array and a tidal baseline period is required, scheduling should begin far in advance of the hydraulic fracture treatment. These logistical considerations and ever-deeper depths of hydrocarbon exploration and production have led to the application of the use of downhole tiltmeters.

1.4.5 Downhole Tiltmeter Mapping. Downhole tiltmeters operate on exactly the same principle as surface tiltmeters but provide significantly different information. **Fig. 1.32** shows a comparison of surface and downhole tilts for the created fracture. As shown, the surface tilt provides information regarding the direction of the fracture (maximum horizontal stress), while the tiltmeters in the wellbore provide information regarding the created fracture height; however, no directional information is provided. As a fracture is created,

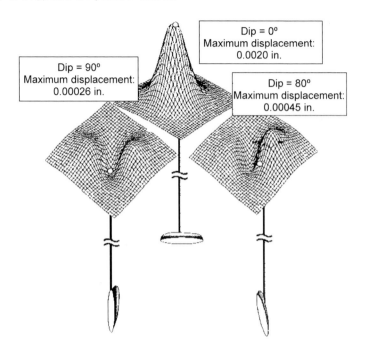

Dip = 0°
Maximum displacement:
0.0020 in.

Dip = 90°
Maximum displacement:
0.00026 in.

Dip = 80°
Maximum displacement:
0.00045 in.

Fig. 1.31—Principle of tiltmeter mapping.

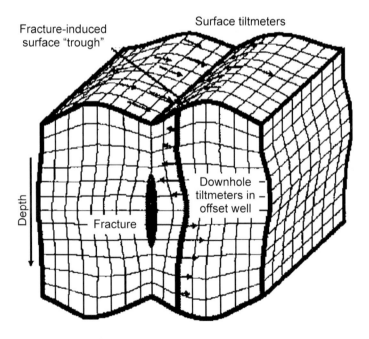

Fracture-induced
surface "trough"

Surface tiltmeters

Depth

Downhole
tiltmeters in
offset well

Fracture

Fig. 1.32—Principle of tiltmeter mapping.

the formation is pushed away from the fracture faces. Directly across from the center of the fracture, no tilt is evident because the motion is purely lateral, while above and below the center of the fracture, the formation in the offset wellbore will tilt in opposite directions (with a reverse nose with the pivot point being the fracture center). When the offset tiltmeter monitor well is close (within a few hundred feet), the top and bottom of the fracture are delineated by the peak tilt magnitude on each of the noses. **Fig. 1.33** shows a theoretical deformation pattern from a downhole tiltmeter located in an offset well 100 ft from the well that is being fracture stimulated. As shown by this depiction, having downhole tiltmeters in close proximity to the well being stimulated can result in quality data for determining fracture height.

1.4.6 Microseismic Mapping. Microseismic theory and mapping has its basis in earthquake seismology. Like earthquakes, microseismic events emit elastic P and S waves (compressional and shear waves, respectively). Although elastic in nature, microseismic events usually occur at much higher frequencies than earthquakes, with a frequency range of 200 to 2,000 Hz. A hydraulic fracture causes an increase in the formation stress proportional to the net fracturing pressure as well as an increase in pore pressure caused by fracturing fluid leakoff. At the fracture tip, large tensile stresses are formed that create significant shear stress. Both the tip processes and pore pressure increases result in shear slippages around the hydraulic fracture. The shear slippages are like mini-earthquakes in the formation with an epicenter within and/or near the hydraulic fracture. **Fig. 1.34** shows a schematic of the process in which the hydraulic fracture is being created and the tip and leakoff processes develop compressional and shear wave arrivals. **Fig. 1.35** shows an example of a three-component geophone and its acoustic interpretation. Using the P and S wave arrival times on the X, Y, and Z components, both location and direction can be determined.

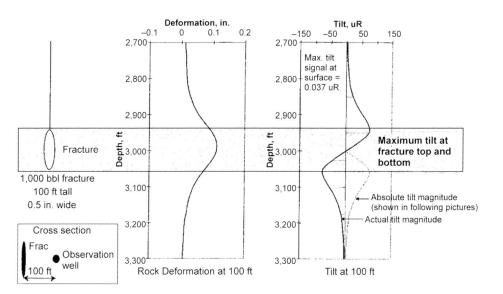

Fig. 1.33—Theoretical downhole tiltmeter deformation pattern.

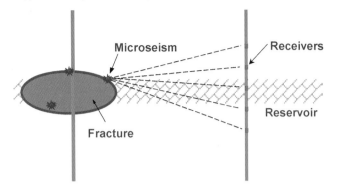

Fig. 1.34—Microseismic-event location. From Fisher (2005).

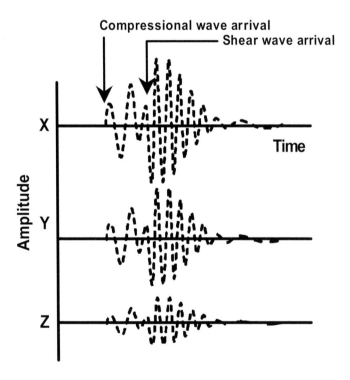

Fig. 1.35—Multiple component geophone array with P and S wave arrivals.

Fig. 1.36 shows a microseismic map of a Taylor fracture stimulation conducted in Carthage Gas Unit No. 21-10. This example will be used in a later section to highlight the benefits of coupling the material balance of fracturing pressure interpretation with the physical measurement of the fracture dimensions (length, height, and direction). As shown, this Taylor sandstone fracture stimulation created a contained fracture (right view) nearly 1,800 ft in half-length (left view).

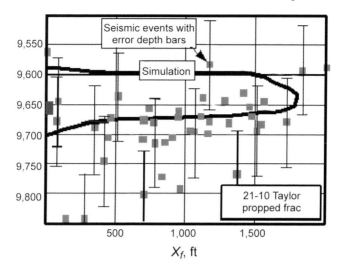

Fig. 1.36—Taylor map.

Besides the breadth of information that is obtained from microseismic mapping, the surveys are much easier to perform than downhole tiltmeter surveys, and the monitor wells can be anywhere from 1,000 to 1,500 ft away. Microseismic mapping has been used in the Barnett shale formation with quite remarkable results. **Fig. 1.37** shows one of the resulting maps that highlight extremely complex behavior from this formation. As shown in this plan view, numerous linear features are evident, possibly indicating the activation of natural fissures or joints in the rock. Note that the linear features are both in the direction of maximum horizontal stress and the natural fissure direction. This behavior has been subject to much debate and scrutiny.

1.4.7 Pressure. Treating pressure and post-fracture pressure decline analysis are some of the most common methods of evaluating fracture treatments. Though surface pressure data are often used, bottomhole pressure measurements are needed to maximize the benefits of this analysis. Furthermore, the use of surface data can lead to erroneous interpretations. **Fig. 1.38** shows a plot comparing the surface and bottomhole pressures from a minifrac pressure decline. Understandably, the surface pressure and bottomhole pressure differed greatly during pumping. However, even after shut-in, in the absence of friction pressure and correcting for hydrostatic pressure, the surface and bottomhole pressures differed both in magnitude and slope. Thus, not only will improper assessments of the fracturing pressure be made, but the material-balance aspects of the minifrac test will be in error as well, as shown by **Figs. 1.39a and 1.39b.** In this real-life example, the surface pressure and bottomhole-pressure-determined fluid efficiencies were 0.52 and 0.29, respectively. This difference in material balance implies a nearly two-fold difference in the required pad fraction: 32% vs. 55%, respectively. For practical purposes, the use of bottomhole data may be limited; however, the engineer should appreciate that not all of the important information is being captured with the surface pressure record. In particular, three pressure tests require the measurement of bottomhole treating pressure (BHTP) whenever possible. These include

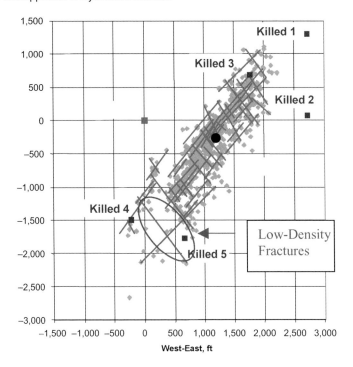

Fig. 1.37—Plan view of fracture-structure plot from one treatment showing the size and complexity of fracture segments in the hydraulic (NE/SW)- and natural (NW/SE)-fracture orientations. From Fisher (2005).

- In-situ closure-stress tests to establish the base fracturing pressure
- Minifrac tests to determine the mechanics of fracture growth (pumping data) and to estimate fluid-loss coefficient (decline data)
- Fracture BHTP analysis to determine the mechanics of fracture growth

In all cases, the pressure data needed to eliminate tubing friction pressure as a factor is the pressure at the perforations. Without bottomhole pressure data, the design engineer is left with matching the final net treating pressure at shut-in and the pressure decline.

Four methods are generally used to measure BHTP (Nolte 1982). One of the methods requires running tubing open-ended (without a packer) and pumping down either the tubing or annulus. The static side (nonpumping side) is then loaded with a fluid of known gradient. Thus, while pumping, the pressures at the surface on the static side, corrected for hydrostatic pressure, are a good measure of the BHTP.

Another method involves the use of a surface readout pressure gauge mounted in a side pocket mandrel, strapping the electric line to the outside of the tubing.

Another technique employs a downhole recording pressure bomb placed into a simple mandrel below a packer. With this technique, actual BHTPs are recorded, but the data cannot be accessed until after the treatment. For the two procedures where BHTP is measured in real time, the stimulation service companies can provide on-site computer vans that facilitate quick manipulation of the prefrac test and/or main treatment data for plotting to make on-site treatment execution decisions.

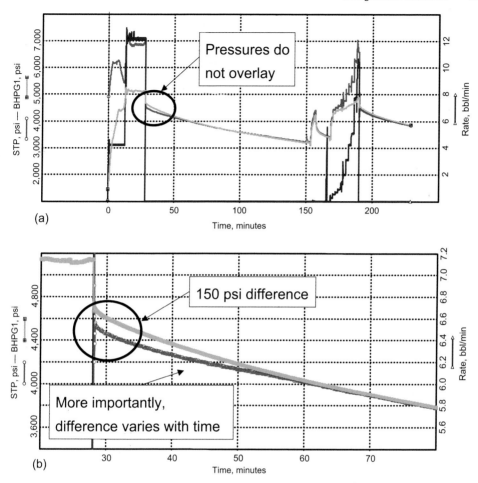

Fig. 1.38—Minifrac pressure decline comparison (surface and bottomhole).

A final method to record downhole pressures includes the use of a casing hanger. The casing hanger and gauge are generally hung in the wellbore above the perforations. This method can allow the gauge to be fixed and the pressure record retrieved without the gauge and hanger being recovered. Following the fracture stimulation, the gauge and hanger are retrieved with a wireline. In recent years, the use of this technique has become more accepted. In some instances, the gauge has been retrieved after the prefrac tests and reinstalled for the fracture treatment while the data were being analyzed.

1.4.8 In-Situ Stress Tests. The most important piece of information to be obtained prior to a stimulation is the closure pressure (minimum in-situ horizontal stress). Closure pressure is measured to determine the minimum pressure necessary to create a fracture, to allow determination of net fracture pressure during a minifrac and fracture stimulation, and to evaluate proppant strength requirements. Even if bottomhole pressure is not measured

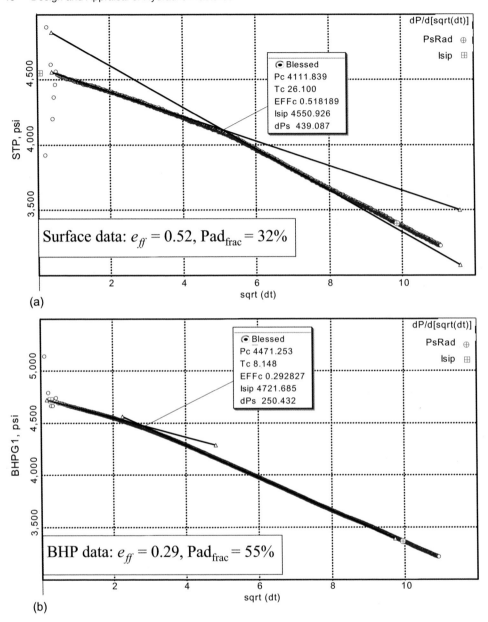

Fig. 1.39—Minifrac pressure decline analysis comparison (surface and bottomhole).

during the treatment (highly recommended at least occasionally), insight into these aspects of the process can be gained by reviewing the ISIPs. This knowledge is of paramount importance to the on-site redesign, execution, and post-appraisal phases.

The subsequent sections will deal with the process of determining fracture closure pressure through the advocation of a wellsite testing program and using it by showing various applications to the treatment redesign, monitoring, and post-appraisal phases.

Fracturing pressure analysis is based entirely on interpreting the "net fracturing pressure" (the BHTP minus the fracture closure pressure or closure stress) of the target formation. The term "closure pressure" is defined as the fluid pressure required to initiate the opening of a fracture. This pressure is equal to, and counteracts, the stress in the rock perpendicular to the fracture.

Closure pressure is equal to or less than the breakdown pressure required to initiate a fracture and less than the pressure required to extend an existing fracture (fracture extension pressure or fracture parting pressure). An upper bound for closure pressure might be estimated from the ISIP after a small volume acid or prepad injection, or from the break-point on a step-rate injection test (fracture parting pressure or fracture extension pressure). However, for quantitative analysis, a more definitive value is needed. While other methods such as logs and core analysis exist to estimate in-situ fracture closure stress, they must all be calibrated by fracturing the rock. Thus, an injection test in which the formation is hydraulically fractured gives the only definitive data for hydraulic fracturing pressure analysis. For measuring closure stress, three basic types of tests are used:

- Pump-in/decline tests
- Pump-in/flowback (PI/FB) tests
- Step-rate injection tests (used to measure fracture extension pressure)

First, let us look at pump-in/decline tests. One such test is the microfrac test. Microfrac tests are a special type of pump-in/decline test used to measure closure stress in a small, discrete zone. The test is usually conducted by perforating a short (1 to 2 ft) interval, typically at 4 to 6 spf, with a 60 or 120° perforation phasing either in permeable formations or in the bounding shales to develop an in-situ stress profile with depth.

These types of stress tests are discussed thoroughly by Warpinski and Teufel (1989) and McLennan and Roegiers (1982). Typically, 0.5 to 1.0 bbl of completion fluid are injected at ⅛ to ¼ bbl/min. The basic theory is that after injecting a small volume (thus the term "microfrac") of low-viscosity fluid at a low rate, the ISIP will be a good approximation of the closure pressure. However, as emphasized by Warpinski (1983), tests must be repeated several times to ensure that the value is reproduced. Generally, multiple repeat tests tend to reduce any influence of the wellbore and rock strength because the fracture is no longer being extended, but only reopened. **Fig. 1.40** emphasizes this point. It should be noted, however, that the ISIP is always an upper bound for closure pressure because a fracture cannot close instantly when pumping is stopped. Therefore, picking an ISIP value and making use of this value must be done with care. When an ISIP is not easily identified, other methods must be used.

The best analysis procedure is to plot pressure vs. the square root of shut-in time. This method is less subjective, and experience suggests that a better estimate is usually obtained. A change in slope indicates a change in the linear flow behavior and is taken to indicate the fracture closing. For example, **Fig. 1.41** shows a cased-hole microfrac stress test conducted in the Mesaverde formation of the Rocky Mountains; clearly no definitive ISIP can be picked. These data were reanalyzed by Miller and Smith (1989) by plotting pressure vs. the square root of shut-in time. As seen in **Fig. 1.42,** agreement between the two analysis methods was nearly perfect.

In addition to microfrac tests, closure stress is determined by pumping a volume of fluid at a rate sufficient to create a fracture, then allowing the fracture to close either by shutting in the well and allowing pressure to decline to below closure pressure or by flowing the

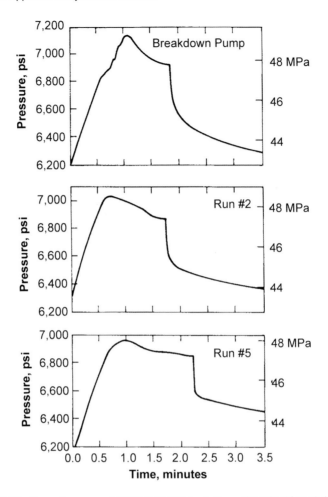

Fig. 1.40—Repeat pumps at 7,531 ft [2,295.4 m]. From Warpinski et al. (1985).

well back until pressure is reduced to below closure pressure (Nolte 1982). In either case, closure pressure is identified by a change in the pressure decline characteristics as the fracture closes. A pump-in/decline test differs from a microfrac test by the volume and pump rate of the injectant. Usually, the base fluid to be used for the fracture stimulation is used for the closure-stress test.

Determination of closure pressure from pump-in/decline tests is operationally very simple. The well is left shut in until pressure declines to a point at which closure pressure can be identified as shown in **Fig. 1.43.** This method of determining closure pressure can be used in high- or low-permeability formations. However, the lower the reservoir perme-ability, the longer the required shut-in time to achieve fracture closure. The data, during a shut-in decline test, should be plotted in real time if possible to determine accurately the length of shut-in time. The decline data can also be plotted on a Horner type plot, **Fig. 1.44,** to identify radial flow and, thus, ensure that the fracture has closed (i.e., lower bound on fracture closure pressure). Also, this plot can be used to estimate the reservoir pressure, p^*.

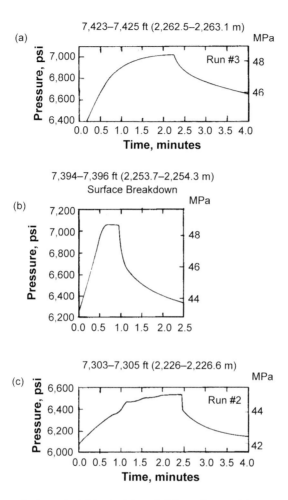

Fig. 1.41—Stress results for paludal tests: (a) coal, (b) sandstone, (c) siltstone. From Warpinski et al. (1985).

Ascertaining the length of shut-in time may require a "trial" test followed by subsequent tests. The number of tests required will depend on the agreement of closure pressures analyzed. If good agreement is evident, only two or three tests may be required. It has been noted that in liquid-filled reservoirs, closure pressure increases with each subsequent test because of an increase in pore pressure. When this occurs, the earlier test results are probably most representative of formation closure and should be used to calculate net pressure during the minifrac and fracture treatment.

Closure-stress determination from flowback pressures is slightly more complicated than a shut-in decline test. However, it may be more effective in low-permeability formations, where extensive pressure decline monitoring is required to achieve fracture closure pressure. A PI/ FB test may be used to accelerate fracture closure, thus making the closure pressure more identifiable. This test is especially useful in extremely low leak-off formations (like shales).

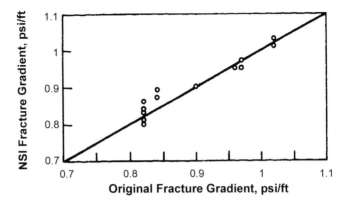

Fig. 1.42—Comparison of stress-test analysis results for MWX, Mesaverde, micro-frac testing. From Miller and Smith (1989).

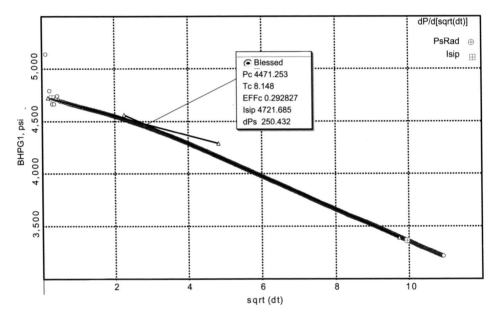

Fig. 1.43—Closure pressure.

For a PI/FB test, the injection is immediately followed by a flowback at a constant rate, typically through a flowback manifold. The constant flowback rate is maintained with an adjustable choke or valve and should be metered with a low-rate flowmeter. The primary purpose of the flowback is to flow back at a rate of about the rate at which fluid is leaking off to the formation. For this flowback rate, a characteristic reverse curvature occurs in the pressure decline at closure pressure. The proper or ideal flowback rate must be determined through trial and error, performing the first flowback at 1 to 2 bbl/min and changing the rate until an "S-shaped" character of the pressure decline is achieved. Once the desired rate is achieved, at least one additional PI/FB test should be performed to ensure repeatability.

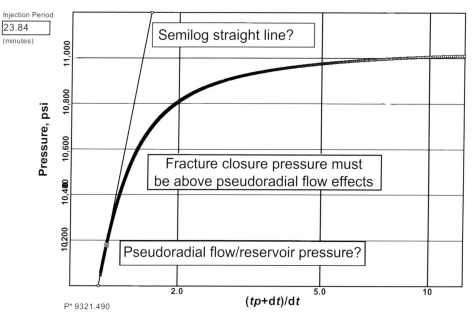

Fig. 1.44—Horner plot.

An additional analysis procedure for PI/FB tests is a derivative plot, such as what is seen in **Fig. 1.45.** For this plot, the change in pressure with respect to time, dp/dt, is plotted vs. time. Because closure is identified at the point at which the rate of decline accelerates, closure would be identified with the maximum point on the derivative plot. For the example in the figure, the derivative is constant for a fairly long period of time (i.e., pressure is declining linearly with time). In such a case, closure probably should be identified at the end of the constant derivative period (i.e., at the point at which the rate of pressure decline begins to accelerate). For this case, it probably would be advisable to run an additional test with a higher flowback rate to achieve a more identifiable maximum on the derivative plot and, thus, a more distinct value for closure pressure. As noted previously, the flowback rate should be held constant. Thus, the PI/FB test requires significantly more operational coordination. Because of this, this test should be used as a last resort for determining fracture closure pressure.

1.4.9 Step-Rate Test. A step-rate test (SRT) to determine extension pressure should always be conducted. The SRT procedure is similar to that performed for reservoir engineering purposes in waterflood operations. Fluid is pumped at constant incrementally increasing rates, and the final injection pressure recorded for each rate is plotted vs. rate as seen in **Fig. 1.46.** A typical test may include rates ranging from 0.25 to 10 bbl/min. The final pressures for each rate step are plotted vs. pump rate, and the breakpoint is identified as fracture extension pressure. For best results, each rate should be maintained for a fixed period of time (typically 1 to 2 minutes). Also, because of the very low rates at the beginning of the test, proper pumping equipment is required (e.g., a low-rate acid injection pump, equipped with a small ID flowmeter). If bottomhole pressure gauges are not used, a reliable SRT can be performed by shutting down after each rate step and obtaining

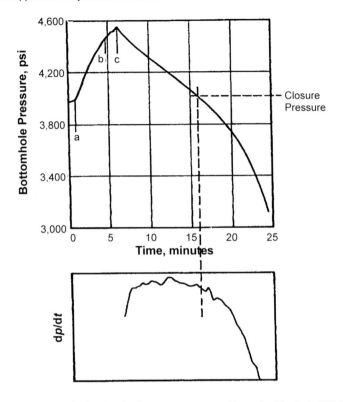

Fig. 1.45—Pump-in flowback closure stress test. From Smith et al. (1989).

an ISIP. To determine fracture extension pressure, plot the ISIP from each rate step as a function of pump rate and identify the inflexion point.

The preceding discussion detailed the tests and analysis procedures that can be used to determine fracture closure pressure. In practice, not all of these tests are required each time to determine the minimum in-situ stress. However, an SRT should be conducted as part of any data collection program. Not only does this test provide a good upper bound on closure pressure (fracture extension pressure), but by monitoring the post-SRT pressure decline, the inflexion as the fracture closes can often be identified. In addition to this test, however, a small injection/decline test should be conducted for confirmation of the SRT results. As a general practice, always ensure that the perforations are broken down prior to conducting an SRT. An example of a data collection program to determine the fracture closure pressure might include

- Loading the hole with completion fluid
- Injecting a small volume (i.e., 1–10 bbl) at a low rate (e.g., 2 bbl/min)
- Shutting in the well and monitoring the pressure decline
- Conducting an SRT
- Shutting in the well and monitoring the pressure decline to fracture closure.

Such a data collection plan provides the ability to determine the fracture extension pressure, reservoir pressure, and two pressure declines to determine fracture closure pressure.

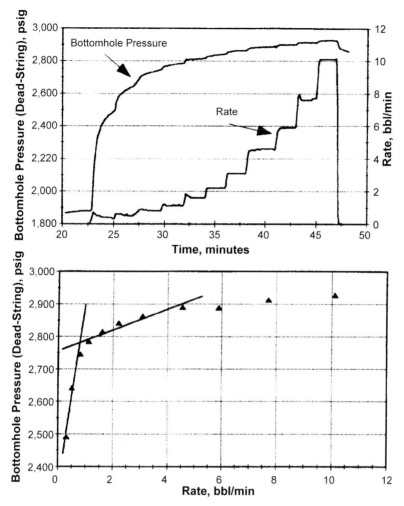

Fig. 1.46—Step-rate test. From Tinker et al. (1997).

If fracture closure pressure is still unclear, additional tests should be considered, given the importance of this parameter to the fracturing process.

Once the fracture closure pressure is determined, it can be used to conduct tests of the fractures material balance in a prefracture data collection program for treatment redesign, to monitor the fracture propagation/geometry during the fracture stimulation, and/or to serve as a post-appraisal tool following the fracture stimulation. To determine the material balance of the fracture stimulation, a minifrac test is recommended. Such a test allows wellsite fracture redesign based on the specific reservoir and fracture parameters from the well of interest.

1.4.10 Minifrac Tests. Before executing a fracture stimulation, a minifrac test should be conducted to refine the geomechanical design parameters. This section of the book will describe this test, its analysis, and its use to redesign or refine the fracture stimulation.

Minifrac tests are pumped to obtain important insight on the mechanics of fracture propagation. These tests, if designed and executed properly, can produce valuable information on fracture geometry, in-situ stress contrast, and fluid-loss coefficient/fracture fluid efficiency. These insights are derived from the test by applying the net treating pressure analysis methods of Nolte and Smith (1981) and the pressure decline analysis techniques of Nolte (1979).

Minifrac tests should be conducted with the planned fracturing fluids (with no proppant) pumped at the planned fracture-stimulation rates. The size or volume of the minifrac should be designed to produce the maximum amount of information about the in-situ stress contrast of the formations of interest. In theory, a minifrac test should be of a size similar to that of the fracturing treatment to fully capture the fracture geomechanics throughout the stimulation. In many formations, however, these insights can be gained from minifrac treatments that are only 10 to 20% of the planned fracture stimulation volume.

Another test used by many in the industry today is the fracture efficiency test (FET) (Shelley and McGowen 1986). This test, though similar to the minifrac test, does not require the use of fracturing fluid to determine the fracturing parameters. Linear (non-crosslinked) gel is often used in this test. Although the interpretations of the minifrac and the FET are conducted similarly, the information gained from these tests can differ greatly. Because the FET is usually conducted with less-viscous fluid, less information pertaining to fracture geometry is obtained compared to a similar-volume minifrac test because the net treating pressure achieved with the FET is probably lower (potentially significantly lower) than that achieved with the minifrac test (reference again Eq. 1.10a). Further, the fracture fluid efficiency/leakoff coefficient determined from an FET may not be representative of that seen during the fracturing treatment when a generally more viscous, better leakoff-control fluid is used. As a result of these differences, FETs should be used only when fracture geometry is fairly simple and known, and where the reservoir controls leakoff. Such an example would be in areas where the fracture is either well or poorly contained (the 2D problem) and the reservoir permeability is extremely low, such as in many tight gas reservoirs.

As mentioned previously, several things can be learned from a minifrac. First, the fracture fluid efficiency can be determined from the analysis of the pressure decline, and second, an estimate of the in-situ stress contrast, modulus, and leakoff coefficient relationship can be determined through a more rigorous history-matching procedure.

Fluid efficiency is defined as the fracture volume (at the end of pumping) divided by the total slurry volume pumped. The term represents a measure of leakoff at the end of the job as one minus the efficiency is equal to the volume of fluid lost to the formation while pumping. In addition, the rate of pressure decline equation can be integrated to determine the volume of fluid lost between shut-in (t_p) and the time at which the fracture closes ($t_p + t_c$). For a minifrac treatment, the volume lost between t_p and $t_p + t_c$ equals the volume of the fracture at t_p. Dividing this volume by the total volume injected gives efficiency. Thus, a relationship between closure time and fluid efficiency exists as shown in **Fig. 1.47.**

The efficiency, e_f, obtained from this figure is used to define a new variable, ρ, which is used in the G-function analysis and defined as

$$\rho V_f / V_L = e_f/(1 - e_f), \quad \dots\dots\dots\dots\dots\dots\dots\dots\dots\dots\dots\dots \quad (1.16)$$

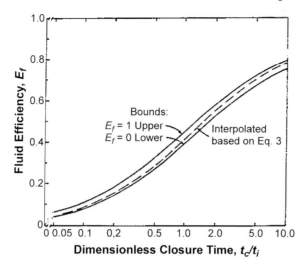

Fig. 1.47—Fluid efficiency vs. fracture closure time. From Nolte (1986a).

where V_f is fracture volume and V_L is fluid loss volume during injection. ρ can also be determined directly from the type curve analysis in terms of the match pressure (Δp^*) and the net fracturing pressure at shut-in (p_s) (e.g., ISIP – closure pressure).

$$\rho = \pi_{ps}/4kG_o\Delta p, \qquad \dotfill \qquad (1.17)$$

where G_o is the pressure difference function at $\delta = 0$ and equal to 1.57–0.238 e_f (within 5%, $G_o = 1.45$), and k is a correction to the fluid-loss coefficient that accounts for additional fluid loss only during pumping (e.g., spurt loss or opening of natural fissures during injection).

Another method to identify nonideal decline behavior is to generate a plot of pressure vs. G-function (Castillo 1987) and dp/dg as a function of the G-function (Barree and Mukherjee 1996), shown in **Fig. 1.48.** As shown, the second derivative decline response would exhibit a straight line (for ideal decline behavior) and would be above or below the ideal line for nonideal decline behavior like pressure dependent leakoff (natural fissures as shown in this figure) or height recession, respectively.

In addition to this quality-control procedure for the decline analysis, a fracture treatment design schedule based solely on fluid efficiency can be generated. Also, efficiency corrections are presented to account for proppant in the fracture at closure, so the pressure decline after an actual propped-fracture treatment can be used in a type curve analysis to calculate fluid-loss coefficient. These modifications to treatment design are found in Nolte (1979).

1.4.11 Net Treating Pressure Monitoring. Bottomhole pressure is the one parameter that can be measured during a fracturing treatment to interpret the fracturing process. All other parameters controlling fracture growth can be related to this pressure. Pressure in the fracture is a function of formation parameters and the fluid system used to create the fracture. If the pertinent rock and fluid properties can be defined, the behavior of BHTP while fracturing can provide valuable insight into fracture growth/geometry characteristics.

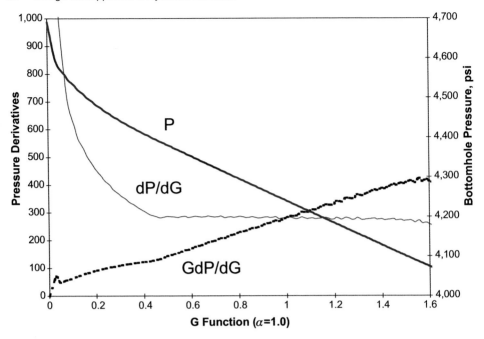

Fig. 1.48—G-function analysis: pressure-dependent leakoff behavior. From Barree and Mukherjee (1996).

The equation used to define net treating pressure was shown as Eq. 1.10, which basically shows that net treating pressure is proportional to the ratio of plain strain modulus to fracture height and the product of fluid viscosity, pump rate, and fracture half-length raised to the quarter-power. This relationship predicts that net pressure should increase with time as fracture length increases, provided that the fracture height is near constant or restricted. If fracture height increases at a constant pump rate and fluid viscosity, there is no reason for the net treating pressure to increase.

However, variations from this prediction of increasing pressure have been observed in numerous cases. The following discussion presents techniques to interpret and analyze these pressure variations to aid in defining the fracturing process for different situations.

Nolte and Smith (1981) developed the net treating pressure analysis methodology used routinely on the wellsite and for post-fracture stimulation analysis. Their log-log plot of net fracturing pressure vs. pump time has proved to be a powerful tool for interpreting the fracturing process. From pressure behavior observations during fracturing, Nolte and Smith (1981) presented a diagnostic framework for fracture monitoring as shown in **Fig. 1.49.** This framework permits the identification of periods of confined-height extension (Mode I), constant height growth (Mode II), restricted extension (Mode III), and uncontrolled height growth (Mode IV). These interpretations are based on combining historical work performed by Perkins & Kern (1961) and Nordgren (1972), showing that net pressure is proportional to time raised to an exponent.

For actual fluids used for fracturing, the exponent (e) can be bounded for cases of high and low fluid loss, and where the fluid's non-Newtonian power law exponent, n, varies from $n = 1.0$ for a Newtonian fluid to $n = 0.5$ for a highly non-Newtonian fluid. For

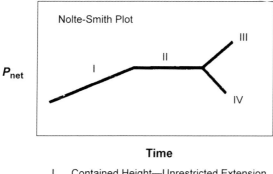

I   Contained Height—Unrestricted Extension
II   Stable Growth/Natural Fracture Opening
III Restricted Extension (Screenout)
IV Unstable Height Growth

Fig. 1.49—Nolte-Smith interpretation guide. From Britt et al. (1994).

the Newtonian fluid with high fluid loss, the exponent e would equal $\frac{1}{8}$. For a highly non-Newtonian fluid with low fluid loss, e would be $\frac{1}{4}$. This defines the boundaries for Mode I fracture extension. As shown in the lower portion of the figure, not only should a contained fracture result in an increase in net treating pressure, but with time, the net pressure should increase with a predictable $\frac{1}{8}$ to $\frac{1}{4}$ slope.

A Mode I, a log-log net pressure to pump time slope of $\frac{1}{8}$ to $\frac{1}{4}$, as discussed above, implies that the fracture is propagating with confined height and unrestricted extension, that fluid loss is linear-flow dominated, and that injection rate and fluid viscosity are reasonably constant. These assumptions comply with the Perkins and Kern fracture growth model.

Fig. 1.50 shows the net treating pressure for three fracture treatments from the Canadian Rockies, DJ basin, and east Texas Cotton Valley, respectively. Note that the initial portion of each treatment indicates confined height and unrestricted fracture extension (Mode I). Beyond this point, however, the treating pressure deviates from the $\frac{1}{8}$ to $\frac{1}{4}$ increasing slope; for the first and third cases in the figure, the slope goes nearly flat, indicating near constant pressure, which characterizes Mode II or critical-pressure behavior.

To analyze what may cause this flattening of the pressure-time slope, the continuity, or mass balance equation, can be examined. The continuity equation simply states that the volume or rate of fluid pumped is equal to the sum of the volume or rate of fluid lost to the formation as leakoff and volume or rate of fluid remaining (stored) in the fracture.

For Mode I behavior, the injection rate and fracture height are constant; fluid loss, fracture length, and net treating pressure increase with time.

In a Mode II environment, a flat pressure-time slope, indicates that the predicted net pressure increase is negated by either stable height growth or increased fluid loss. The potential for height increase is shown in **Fig. 1.51,** wherein the fracture penetrates a section of higher stress at a constant growth rate. As additional height is generated, the cross-sectional area of the fracture increases, thus reducing the flow velocity and frictional pressure drop down the fracture and reducing the normal pressure increase. If height growth continues and reaches

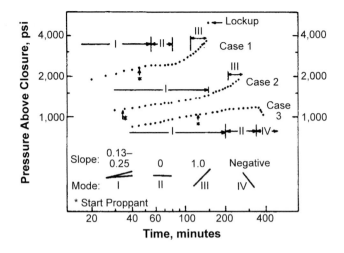

Fig. 1.50—Example of different characteristic slopes. From Nolte and Smith (1981).

a low-stress zone, as seen in the figure, the pressure-time slope may become negative, indicating uncontrolled, rapid height growth (Mode IV). This type of behavior is discussed later.

The other variable that can change besides height without violating the continuity equation is fluid loss. One mechanism for a higher fluid-loss rate would be opening of natural fissures intersected by the main fracture as shown in **Fig. 1.52.** The opening of natural fissures increases fracture volume and fluid-loss area, and it decreases the pressure in the fracture. When pressure declines below the stress holding the fissures closed, the fissures close up again. Pressure then increases slightly and the fissures reopen. This opening-closing-opening of the fissures is like a pressure regulator, producing a constant pressure profile. Because of the increased fluid-loss rate as well as the near-wellbore location of the increased fluid loss, Mode II will normally lead to a premature screenout of the treatment.

If something occurs to stop fracture extension, then either net treating pressure or height must increase. Increased fluid loss to natural fissures may dehydrate the slurry to the point that a proppant bridge forms in the fracture. If pumping continues, no additional fracture penetration will occur. In a contained fracture, pressure must increase at a higher rate, as seen in Cases 1 and 2 on Fig. 1.50. If the fracture is not contained, the rate of height growth will increase and pressure will decrease with time, as shown by Case 3 of Fig. 1.50. In the case where the fracture is contained and the pressure increases, this rapid pressure increase is characteristic of Mode III behavior on the log-log Nolte-Smith plot, as shown in Fig. 1.50.

Mode III is characterized by a region of positive unit slope (i.e., 1:1 log-log slope), indicating a flow restriction in the fracture. This implies that the pressure is proportional to time or, more importantly, that the incremental pressure change is proportional to the incremental injected fluid volume. This 1:1 slope is similar to the unit slope in pressure transient analysis, indicating storage of fluid—in this case, by swelling or ballooning the fracture. Common causes of this behavior are pad depletion (where proppant reaches the fracture

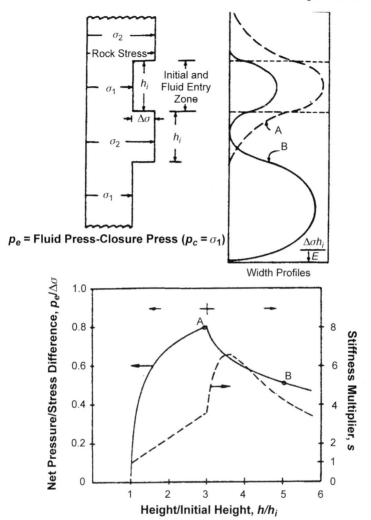

p_e = Fluid Press-Closure Press ($p_c = \sigma_1$)

Width Profiles

Fig. 1.51—Height vs. net pressure. From Nolte (1988).

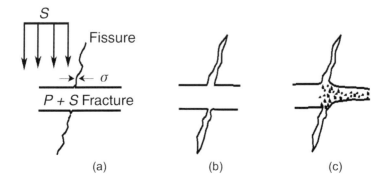

Fig. 1.52—Natural fissure opening schematic.

tip), slurry dehydration to natural fissures, excessive height growth increasing fluid-loss area, and/or proppant fallout associated with poor gel quality.

Fig. 1.53 shows how excessive height growth can cause slurry depletion, resulting in a premature screenout. The fracture has grown through a shale section into a lower-closure-pressure sand. Because of the higher stress in the shale, the fracture width is less than that in the sands, forming a "pinch point" that will not allow sand to pass through, yet that allows fluid to pass, dehydrating the slurry in the target interval. As the slurry dehydrates, it forms a plug that will eventually bridge in the fracture. The approximate distance to the bridge can be calculated from

$$R_{max} = 1.8 \ qE'/H^2 \ (\partial p/\partial t), \ \dotfill \ (1.18)$$

where q = pump rate (bbl/min), E' = plain strain modulus (psi), H = frac height (ft), and $\partial p/\partial t$ = rate of pressure increase (psi/min). This information can be useful in post-analysis and in the design of future treatments. A near-wellbore bridge would probably be caused by natural fissures, height growth, or a high-sand-concentration slug; whereas a bridge some distance from the wellbore would more likely be caused by pad depletion or by sand fallout associated with poor gel quality.

As noted previously, if fracture extension ceases and the fracture is not contained, then rapid, unstable height growth will occur as pumping continues, and the pressure-time slope will become negative. This is Mode IV behavior as seen during Case 3, Fig. 1.50.

In Mode IV, a negative slope can be interpreted as rapid height growth into a lower-closure-stress zone. Referring back to the continuity equation, a significant decrease in pressure must be accompanied by a significant increase in one or more of the other variables. A significant increase in fluid loss is possible from opening new fractures or fissures, but it is

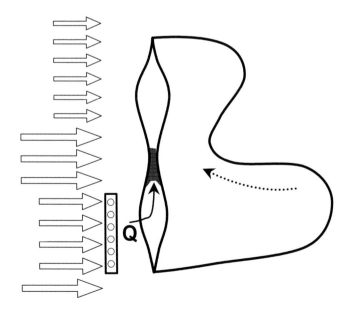

Fig. 1.53—Height vs. net treating pressure.

not likely with decreasing pressure. An increase in length is not consistent with a decrease in pressure. The only change compatible with a decrease in pressure is an increase in height.

The steepness of the negative slope would imply the rate of unstable growth. A high rate of growth would exhibit a steep slope, while a low negative slope would imply a low rate of growth. If the fracture grows into a much-lower-stress zone, the decrease in pressure will be rapid. If the fracture grows into a slightly-lower-stress zone, the negative slope will be shallower.

A negative slope observed from the beginning of the treatment indicates a lack of height-confinement. In this case, the fracture will grow radially, and future treatments should be designed using a radial model.

While the observed pressure behavior on the net pressure vs. time plot is primarily a function of fracture geometry, other parameters may interfere with interpretation. These parameters include the pump rate and fluid viscosity. Although each of these parameters has only a quarter- power effect on the net treating pressure, large variations in either parameter can have an effect on fracture net pressure monitoring.

As mentioned previously, Mode II behavior on the net pressure vs. time plot is usually followed by some undesirable behavior such as excessive height growth or a screenout. For this reason, the net pressure at which the pressure-time slope flattens is termed the *critical pressure*. For the case of height growth, critical pressure is roughly 70 to 80% of the differential closure stress between the initial zone and bounding beds. When natural fissures exist, critical pressure is approximately the net stress component (above closure pressure) acting normal to the plane of the fissures, holding the fissures closed.

In fieldwide studies, critical pressure has been found to be reasonably constant. During the early development of a field, strategic wells should be monitored to determine the critical pressure, which then can be extrapolated to offset wells. Treatment designs then can be formulated to keep net treating pressure below the critical pressure, possibly by reducing viscosity or rate. If it is impossible to stay below critical pressure by these means, unconventional-type designs may be developed to minimize height growth or screenout tendencies.

To perform a meaningful analysis of fracturing pressures requires direct measurement or a very accurate calculation of BHTP during the injection and pressure decline. The primary objective is to record the fluid pressure at the entrance to the fracture (e.g., just outside the perforations). While several companies have developed software for calculating BHTP from surface pressure, to date no technique has been developed to accurately account for all variables affecting friction pressures. In some cases of shallow wells, where injection was down large casing, these software programs have given reasonable results. In deeper wells, however, and especially those where the fracture treatment was pumped down tubing, results have been erroneous, and in many cases have led to incorrect decision making during the treatment.

Although bottomhole pressure data are required to do rigorous net-treating-pressure analysis of fracture behavior, these techniques can be useful and can provide insight into the fracturing process for the multitude of treatments pumped without bottomhole pressure gauges. For example, the post-fracture ISIP can provide information regarding fracture geometry, assuming fracture closure pressure is known. Applying the continuity equation as before in a hard-rock application, if the final net treating pressure is low, the fracture geometry is likely to be radial (with significant height growth). If the final net pressure is high, the fracture geometry is likely to be contained (with significant fracture extension). Of course, an impending screenout (tip event) could result in a high post-fracture ISIP and elevated final net pressure, negating such a simple interpretation. Further, in many cases the final net pressure is intermediate, thus providing no insight into the fracture propagation and geometry that occurred prior to shutdown. Even in these cases, additional insight can be gained if during the pad the pumps are shut down and an ISIP is obtained. After the treatment, the ISIPs can be compared, and fracture geometry probably can be ascertained. If prefrac testing is conducted (and this is highly recommended), the ISIPs from the tests can be compared to provide similar insight.

1.5 Treatment Execution and Monitoring

1.5.1 Quality Control and Assurance. Quality control and assurance is one of the most important and underappreciated components of fracture stimulation execution and post-fracture appraisal. Good quality assurance provides the engineer with the ability to design, monitor, predict, and determine the achieved fracture dimensions. All benefit from a good quality-assurance program. For success to be achieved, everyone must ensure that what was supposed to be pumped was in fact pumped. In its most basic form, this assurance program should include an understanding of the fracturing materials and components prior to wellsite execution and the material-balance aspects of the fracturing materials (fluid, additives, and proppant) both before and after the stimulation. It should also include collection of wellsite samples during the stimulation and review of samples afterward. Any quality-assurance program also should include strict adherence to safety guidelines.

1.5.2 Coupling of Pressure Analysis and Microseismic Mapping. The application of pressure analysis to hydraulic fracturing is a tremendous aid to understanding the process. However, while pressure analysis is a powerful tool for routine application, pressure analysis has two major limitations. First, the measured pressure at the wellbore represents a final result of an entire process and is therefore controlled by many variables; second, the final interpretation of pressure data for specific fracture dimensions such as length or width must come from a "model," leading to more uncertainty and possibly differing results. The many variables and models were discussed in some detail earlier in the book. However, pressure analysis becomes more definitive when the treating pressure analysis is combined with pressure decline analysis. With the pressure decline analysis, the pressure analysis becomes a powerful "material balance" tool. Much as reservoir engineers use reservoir pressure to define the material balance for producing fields, fracturing-pressure-decline analysis provides good values for the overall material balance (i.e., "volumetrics") of the fracture, but without information on specific dimensions.

While pressure analysis is a powerful tool, clearly definitive answers demand something additional, and many other fracture diagnostics have been used. Possibly the most common is the use of post-frac radioactive logs to measure fracture height. Such gamma ray logs are a good tool and an extremely powerful tool when long perforated intervals and/or multiple perforated intervals are involved with the fracture treatment. However, where there is (or may be) extensive height growth above/below the perforated interval, this diagnostic tool becomes less certain. In addition, it can measure fracture height only at or very near the wellbore. As a result, these post-fracture diagnostic logs are better suited for cases in which the fracture is likely to follow the wellbore, such as where there is strong directionality in fracture behavior ($\sigma_{Hmax} >>> \sigma_{hmin}$).

Fracture imaging through microseismic monitoring (Warpinski 1983; Warpinski and Teufel 1989), on the other hand, yields good values for specific dimensions of fracture length and height, but without the material balance (fluid loss and width) information needed for treatment design. In addition, on the negative side, microseismic imaging is more difficult to measure, and except for unusual conditions, it would not be available for routine application. However, the two procedures (fracture pressure analysis and microseismic imaging) clearly complement one another. Combining seismic imaging with the more usual pressure-analysis-based design procedures should allow complete fracture optimization/design founded on a firm basis. Subsequent sections will show the application of both pressure analysis and microseismic imaging to the Taylor sand in the east Texas Cotton Valley field.

The first part of the normal fracture design process begins with determining the major design variables or geomechanical profile with depth. The major design variables include in-situ stress, Young's modulus, and fluid-loss coefficient.

While processes such as modulus contrast, or "interface slip" (particularly in highly over-pressured reservoirs), may sometimes, in some places, play roles in fracture height confinement, in-situ stresses are normally the dominant parameters in controlling height (Miller and Smith 1989). At a minimum, in-situ stresses always determine maximum height for a fracture. That is, other processes may limit height growth, but height can never be greater than that allowed by in-situ stress differences [as expressed in the original work of Simonson et al. (1978)]. In fact, microseismic imaging offers a diagnostic for determining if/when such "other processes" may be playing a role.

The in-situ stress data used in this example were derived from a dipole sonic log interpretation and calibrated with in-situ stress tests of individual zones. The log prediction was calibrated against historical, published (Veatch 1983a) stress data from the Cotton Valley formation. The validity of this historical correlation was then tested against more recent in-situ stress tests conducted in conjunction with the microseismic imaging project. These stress tests confirmed the calibrated stress profile, as shown in **Fig. 1.54.**

The modulus of the formation was then determined through triaxial compression testing of the core in the laboratory, and a modulus profile with depth was developed by correlating the laboratory static measurements with the dynamic measurements from the dipole sonic log.

The fluid-loss coefficient for this tight formation gas reservoir was expected to be on the order of 0.0005 to 0.001 ft/min$^{1/2}$ on the basis of estimates of the filtrate, reservoir, and wall effects, but it was determined in-situ from prefrac testing as described previously.

To test or validate the geomechanical profile with depth, a hydraulic fracturing simulator was used to analyze the treating pressure and pressure decline data both from the prefrac-data collection and during the actual fracture stimulation.

Fig. 1.55 shows a summary of the prefracture tests and actual fracture treatment. As shown, SRTs and minifrac tests were conducted prior to the fracture stimulation. Next, the minifrac treatment was history-matched as shown in **Fig. 1.56,** a plot of net treating pressure vs. pump time. Note that for the analysis, the dead string pressure during the minifrac treatment was converted to bottomhole pressure by adding a hydrostatic head of 4,250 psi and subtracting the closure pressure of 6,290 psi and 140 psi of downhole friction to generate the net pressure plot of the minifrac test. These data then were plotted and matched as shown in the figure. This match of the 30,000-gal treated-water minifrac test required only a slight modification to the geomechanical data established previously because the leakoff coefficient had to be increased to 0.0012 ft/min$^{1/2}$ (preliminary estimates were 0.0005–0.0010 ft/min$^{1/2}$).

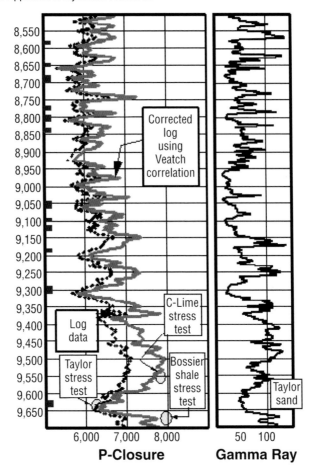

Fig. 1.54—Stresses.

Next, the actual fracture stimulation treatment pumped in the Taylor sand (100,000 gal of low-guar crosslinked gel with 162,000 lbm of 20/40 Ottawa sand) was modeled with the refined geomechanical model used to match the minifrac test. A final net treating pressure of 1,300 psi was seen during the fracture stimulation, with a tip screenout probably occurring at the end of the job. As shown in **Fig. 1.57,** a comparison of the model and actual net treating pressure data, the model correctly predicted the net pressure gain, with predicted dimensions of a fracture half-length of 1,800 ft, a fracture height of 110 ft, and an average conductivity of 220 md-ft (0.7 lbm/ft² average). This match of the pressure data in the Taylor zone of the Cotton Valley formation represents the material-balance match of the fracturing process. Next, the microseismic image was reviewed to determine the fracture dimensions.

The monitor wells in this field example were placed to monitor the east wing propagation of the fracture. The microseismic data collected in the Taylor sand showed an east wing fracture half-length slightly greater than 2,000 ft. The fracture height noted from the microseismic data yielded a fracture top of 9,562 ft and a fracture bottom of 9,842 ft; thus,

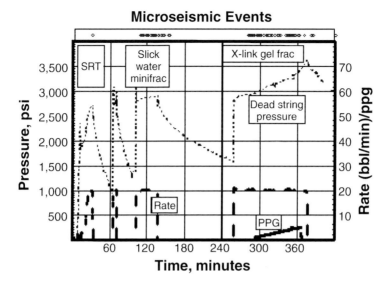

Fig. 1.55—Tests.

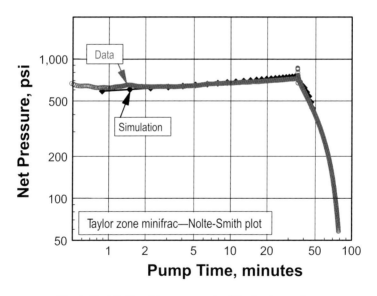

Fig. 1.56—Minifrac history match of Taylor formation.

the fracture height was equal to 280 ft. Comparison of the microseismic data was made to the bottomhole-pressure-predicted fracture behavior. A Nolte-Smith plot of this data is included in **Fig. 1.58.** After the viscous gel is on the formation for about 40 minutes, a positive ¼ slope indicates confined height fracture extension, with viscous forces dominating fracture behavior. Then, at a time of about 85 minutes, a net pressure unit slope

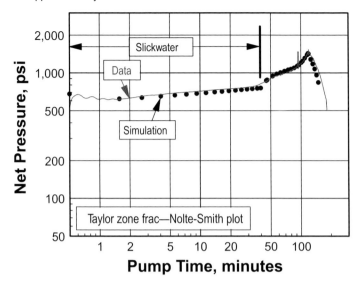

Fig. 1.57—Fracture stimulation history match of the net pressure.

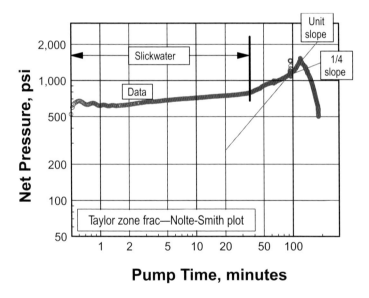

Fig. 1.58—Fracture stimulation history match.

develops, indicating that some form of restriction has formed in the fracture and that, presumably, fracture extension has stopped.

This seems to compare reasonably well with the seismic image (**Fig. 1.59),** which shows very little growth occurring after about 90 minutes. This figure also shows that the model-predicted results and the seismic image of x_f vs. time are in quite reasonable agreement throughout the entire treatment. The only discrepancy is the final seismic event at a fracture

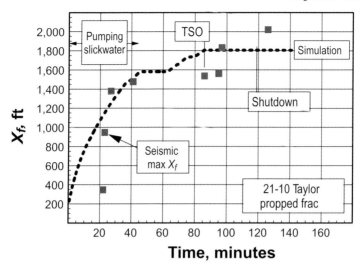

Fig. 1.59—Comparison of microseismic image and fracture model.

half-length of 2,021 ft, and this event actually occurred after the well was shut in. Further, both the treating pressure and the microseismic interpretation suggest that the height growth of the fracture probably occurred late in the treatment. This growth into the lower barriers resulted in a narrow fracture with little impact on the material balance of the process. Interestingly, this height growth occurred at a net treating pressure of 1,200 psi. This net pressure is consistent with Case 2 of Fig. 1.57 (also a Taylor sand example) in which a critical pressure of 1,200 psi was reached and fracture height growth occurred.

Chapter 2

Post-Appraisal

2.1 Introduction to Post-Appraisal With Dynamic Data

After designing and placing a fracture stimulation, it is desirable to establish how close the actual fracture is to that designed and expected. For instance, things like proppant crushing, proppant settling, poor fluid cleanup, ineffective breaker action, or closure of unpropped sections can affect the in-place fracture. And, of course, the design process itself is based on a simple picture of the reservoir and is driven by simplified models of the fracturing process. All of this leads to the conclusion that the actual fracture stimulation is probably not going to be identical to the designed fracture stimulation.

Ideally, we would like to find a way to evaluate the effective properties of the in-place fracture. Though post-appraisal starts with simple material balance calculations on pumped pad and proppant volumes (measures of created and propped fracture volume), tracking of recovered fluid volumes and composition (measures of cleanup) and metering or estimation of early hydrocarbon production rates (indicators of successful stimulation), quantitative characterization relies almost exclusively on the collection and analysis of dynamic data.

Collecting dynamic data involves measuring the *bottomhole pressure changes in response to an imposed and controlled change in production or injection rate*. The prefracturing data described in Chapter 1 (SRTs, minifracs, and other pump-in tests) are all examples of dynamic data (the results of which were used in designing the stimulation).

The dominant dynamic data sets (pressure and rate vs. time) collected during a fracture stimulation are shown in **Fig. 2.1.** In the context of common industry process, these data sets are related to the plan + do + measure + learn cycle. Part A of this figure shows prefrac data

---

Section Takeaway

- Post-appraisal lies in the "measure + learn" part of the completions cycle.
- Classic or traditional post-appraisal begins with post-fracture stimulation "flow and buildup" tests.
- Analysis of flow and buildup data fit within the framework of pressure transient analysis theory.
- Long-term performance data represent an "uncontrolled" post-appraisal data set.
- The rate-time analysis method can be used to extract information from long-term performance data.

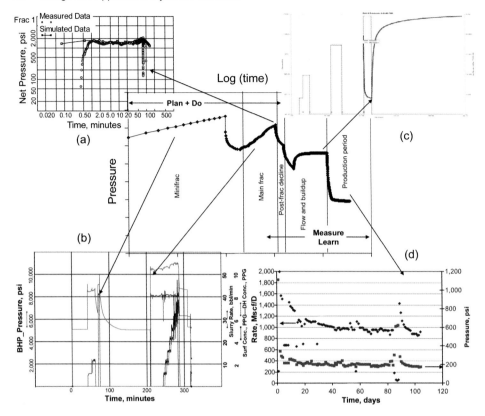

Fig. 2.1—Pressure response periods during a fracture stimulation.

collection and the main fracture stimulation period. These periods correlate with the plan + do step in the fracture stimulation completions cycle and were the focus of Chapter 1. An analysis of these kinds of data is illustrated in Part B of this figure, showing a fracture simulator match of the net-treating pressure. Both of these data sets are typically a few tens of minutes to a few hours in duration.

The third dynamic data set, shown as Part C of Fig. 2.1, represents the traditional beginning of the post-appraisal period and shows the most common data set collected for the post-appraisal of a well completion. This part of the figure illustrates a flow and shut-in (pressure buildup) test. Flow, for our purposes, is defined as the time trace of the pressure in response to producing the well according to a controlled rate sequence, while buildup is the time trace of the pressure in response to closing in the well (instantly reducing the production rate to zero) after the flowing period. Both data sets are typically tens of hours to a few days in duration. These data sets lie in the measure + learn part of the fracture stimulation completions cycle and fit within the analysis framework of classic pressure transient analysis. Because this is a highly developed theoretical framework, the analysis of these data sets is discussed first and forms the topic of this section.

The final dynamic data set is shown as Part D of Fig. 2.1. This data set is an exception to the definition of dynamic data given above in two ways. First, the changes in rate are not imposed and controlled, but rather are the result of any number of operational or well-condition issues. For example, in this data set, the discontinuity in the rate near the end of the plot is due to an unplanned shut-in of production from the well. Another common unplanned change in well rate occurs when the well rate falls below the values needed to continuously carry associated water and condensate to the surface. When the rate falls below this level, liquids accumulate in the well, the backpressure on the sandface increases, and the well stops flowing (logs off). After a sufficient period, the pressure in the well increases enough to allow it to flow at high rates again, temporarily offloading the liquids. Other decreases or increases in rate that can occur are caused by backpressure changes resulting from changes in sales line (surface) pressure. Second, the actual values of rates are not instantaneous measurements but, rather, are average daily values (i.e., the rates reflect daily cumulative production, while the pressures are instantaneous values reported from the automation system at a generally consistent time during the production day). The difficulty with the use of daily cumulative production as a substitute for instantaneous rate is that changes in pressure from day to day are not a direct response to the changes in daily rate. For example, the well pressure could be reported at the end of a short shut-in period, even though the well had produced for the majority of the day.

Nevertheless, the data set shown in Part D of Fig. 2.1 can be extremely useful for longer-term learning. These data have advantages. For one, they are almost always available. Also, as is obvious from the plots in Fig. 2.1, this data set is by far the longest in duration. In general, the length of this data set is more of an advantage in prediction of rate and reserves than in stimulation post-appraisal. However, the length of this data set offers an advantage for judging the effectiveness of stimulations performed on wells in low-permeability formations because, as we shall see, wells in these so-called "tight" (low-permeability) formations require prohibitively long pressure-buildup test times.

The data set in Part D fits into the theoretical framework of decline curve or rate-decline analysis (RDA). RDA methods have their basis in classic PTA and have been a major area of active research and development over the last 15 years. Though this form of analysis can be and often is used in stimulation post-appraisal, the approach is historically lumped with reserves-estimation and rate-prediction methods. For this reason, we will delay the discussion of this data set until Chapter 3 of this book, where we will use it mainly in a performance prediction context.

2.2 PTA

PTA is a reservoir engineering technology with its roots in similar methods used in groundwater hydrology. These methods have played an established role in hydrocarbon well-formation evaluation since the late 1940s, with the publication of work by van Everdingen and Hurst (1949). PTA fits into the branch of applied mathematics called inverse problem theory, in which mathematics addresses the estimation of the properties of a system through the analysis of system output induced by a specified input (Tarantola 1987).

Section Takeaway

- PTA is a petroleum engineering discipline fitting into the general area of inverse problems.
- Many analysis procedures in petroleum engineering also are inverse problems.
- PTA methods are based in analytical solutions derived for approximate models of the actual physical systems.

PTA, like all inverse problem solution approaches, uses the following steps. First, the physical system is reduced to a conceptual system of reduced complexity. Second, a set of equations is developed relating the output of the conceptual system to the form and strength of the input signal and the values of the conceptual system parameters; in most instances, the input signal is made as simple as possible. Third, this set of equations is solved to yield expressions for computing the values of system output induced by the simple input signal.

To illustrate this process in the context of PTA, we will briefly review the basic single-rate flow problem consider by Theis (1935). Not only will this illustrate the process, but it will provide one of the primary results we will use throughout the remainder of Chapter 2.

2.3 Drawdown Paradigm and the Radial Flow Regime

Consider the following scenario. A new well is drilled into a formation of unknown quality. The well penetrates an interval where evaluation logs indicate that the formation at the well contains exploitable hydrocarbons. The basic question we would like to answer is this: If the well is produced, can the measured pressure response be analyzed in some way to obtain some additional information about formation characteristics?

We reduce this physical system to the conceptual system shown in **Fig. 2.2.** The main system-simplifying assumptions are these:

- The formation contains only a liquid hydrocarbon phase, and the well is above its bubblepoint pressure, with a constant and "small" compressibility. In addition, the viscosity of this liquid hydrocarbon phase is constant.
- Any water present is immobile, with constant compressibility.
- The formation is at virgin conditions, and its pressure is constant throughout.
- A cylindrical well fully penetrates the formation, and there is perfect connectivity between the formation and well.
- The formation has constant properties (thickness, porosity, permeability, and compressibility) and extends to infinity in all directions.

We will also assume that the well is produced at a constant rate, measured at surface. This implies that the bottomhole rate is related to the surface rate through the formation volume factor, B, which we will also assume to be constant.

With this definition of the conceptual system, we now can proceed to develop the equations relating the output of this conceptual system to the form and strength of the input signal and the parameter values of the conceptual system. In this case, as is usually the case in pressure transient analysis, the output signal is the measured change in flowing bottomhole pressure as a function of time, $p_{wf}(t)$, while the input signal is the imposed production of the well at a constant surface rate. Because the well is a cylinder that fully penetrates the

---

Section Takeaway
- The drawdown paradigm is the basis for all PTA methods .
- All PTA models have a radial flow regime as part of the ideal model response.
- Permeability can be estimated from this flow regime.

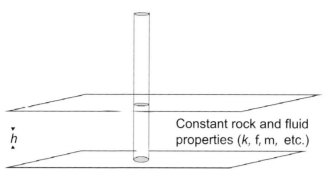

Fig. 2.2—Theis conceptual model.

reservoir, the flow to the well will have radial symmetry, and the isobars (contours of constant pressure) in the reservoir will be cylinders concentric with the well. Thus, the pressure at any point in the reservoir is a function of radial distance from the well and time, $p(r, t)$. Combining the continuity equation (conservation of mass) with Darcy's law leads to a partial differential equation governing $p(r, t)$. When this equation is solved subject to appropriate boundary conditions (Raghavan 1993), we obtain

$$p_i - p(r, t) = -\frac{qB\mu}{4\pi kh} Ei\left(-\frac{r^2}{4\eta t}\right). \qquad (2.1)$$

This is the solution describing the way pressure changes with both time and radial distance in the conceptual system of Fig. 2.2. To arrive at this solution, we have had to further assume that the well volume has been reduced to zero. This is usually referred to as the line-source well assumption. Evaluating this solution at $r = r_w$ gives the predicted behavior of the flowing bottomhole pressure, $p_{wf}(t)$, as

$$p_i - p_{wf}(t) = -\frac{qB\mu}{4\pi kh} Ei\left(-\frac{r_w^2}{4\eta t}\right). \qquad (2.2)$$

The symbol $-Ei(-x)$ is a special function called the *exponential integral*.

Eq. 2.2, the solution presented by Theis for a constant-rate problem, is the expression that can be used to analyze measured values of flowing bottomhole pressure to estimate k and $\phi\mu c_t$ via type curve matching (Agarwal et al. 1970).

Eq. 2.2 is cumbersome in practice when trying to estimate system parameters. The common approach to overcome this in PTA is to develop simple linear approximations to the full solution that are valid over limited time periods. These limited time periods are called *flow periods* or *flow regimes*. Though generally a solution may have many such flow regimes, the Theis solution has only one.

To develop the flow regime associated with the Theis solution, we first note that for small, strictly positive values of ε, we can approximate the Ei-function by (Dake 1978)

$$Ei(-\varepsilon) = \ln(\varepsilon) + \gamma, \qquad (2.3)$$

where γ is Euler's constant. Using Eq. 2.3 in Eq. 2.2 gives

$$p_i - p_{wf}(t) = -\frac{qB\mu}{4\pi kh}\left[\ln\left(\frac{r_w^2}{4\eta t}\right) + \gamma\right]$$

$$= \frac{162.6qB\mu}{kh}\left[\log t + \log\frac{k}{\phi\mu c_t r_w^2} - 3.23\right], \quad \text{.................} \quad (2.4)$$

where we have obtained the last equality in Eq. 2.4 by changing to standard field units and replacing the natural logarithm function with the base-10 logarithm.

Eq. 2.4 defines the *radial flow regime*. This flow regime is very important throughout pressure transient analysis because it occurs in every analytical model.

Eq. 2.4 shows that a plot of pressure drop, $p_i - p_{wf}(t)$ vs. the logarithm of producing time will give a straight line with slope, m, inversely proportional to the permeability-thickness product, kh. Specifically,

$$kh = \frac{162.6qB\mu}{m} \quad \text{..} \quad (2.5)$$

Fig. 2.3 illustrates the analysis procedure outlined in the above paragraph and in Eq. 2.5. Entering this figure at 10 hours and 100 hours and reading the pressures from the drawn straightline trend, we obtain a value for m of approximately 534 psi/log cycle. Using the known values of $q = 500$ B/D, $B = 1.15$ RB/STB, $\mu = 0.7$ cp, and $h = 25$ ft, we calculate

$$k = \frac{(162.6)(500)(1.18)(0.7)}{(534)(25)} = 4.9 \text{ md.}$$

Fig. 2.3 illustrates the analysis procedure outlined in the above paragraph and continued in Eq. 2.5. The reader should note that choosing the correct region in the data for constructing the radial flow period straight line is critical to this analysis. The general problem of flow period identification is a central one, and we will discuss it later in this section.

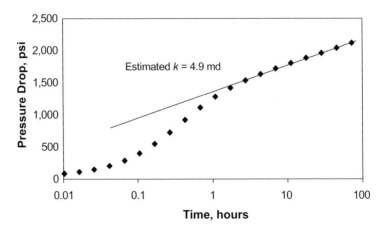

Fig. 2.3—Simulated example radial flow regime plot (input $k = 5$ md).

2.4 Skin, Wellbore Storage, and Boundary-Dominated Flow

The three most limiting assumptions of the Theis solution are

1. The assumption of perfect connectivity between well and formation
2. The assumption of zero wellbore size and volume (line-source well)
3. The assumption of the infinite extent of the reservoir

We will now discuss how these assumptions can be relaxed to obtain some more general expressions.

2.4.1 Van Everdingen and Hurst Skin. The simplest, most common, and most important problem affecting well-to-reservoir connectivity is that induced by a reduction or enhancement of the ability of the hydrocarbon fluid to move through the near-wellbore region and into the well. Reduction in connectivity is equivalent to damage, and enhancement is equivalent to stimulation. Some damage mechanisms are

- Invasion of drilling mud-filtrate
- Poor perforating efficiency or failure to perforate sections of the pay zone
- Inertia-turbulent or non-Darcy flow [i.e., the pressure drop is no longer linearly related to the rate as it is in Darcy's law (non-Darcy flow does not mean turbulent flow in the classic Reynolds number sense)].
- Inefficient cleanup of completion fluids, especially where lost circulation materials are required
- Poor removal of mudcake in openhole gravel-pack completions

Some mechanisms that yield a stimulated condition are

- Pumping acids
- Deep-penetrating perforating methods
- Formation breakdowns
- Hydraulic fracturing

All near-well damage or stimulation mechanisms are incorporated into the Theis solution through the van Everdingen and Hurst skin effect. In this approach, there are two important ideas to understand. First, damage and stimulation are assumed to result in pressure changes that stabilize essentially instantaneously with any change in rate. Second, damage or stimulation is defined by a difference in measured pressure relative to that predicted by the Theis solution given in Eq. 2.2. These assumptions have their foundations in the observed behavior of producing wells. Damage (positive skin) is taken to be any

---

Section Takeaway

- Skin is a measure of near-wellbore damage or stimulation.
- Skin is quantified relative to the ideal well of the Theis solution.
- Skin has a number of physical sources such as fluid invasion and hydraulic fracturing.
- Fracture stimulation can mitigate a number of sources of positive skin.
- Relative to the ideal well, fracturing generally creates negative skin.
- Skin is estimated from the radial-flow-regime data.

constant, stable decrease in flowing bottomhole pressure when compared to the bottomhole pressure predicted by the Theis solution. Similarly, stimulation (negative skin) is any increase in bottomhole pressure in comparison with this reference. Formally, we can write

$$\Delta p_{skin} = p_{wf,Theis}(t) - p_{wf,s \neq 0}(t). \qquad (2.6)$$

This equation should be thought of as the way to predict the flowing bottomhole pressure influenced by damage or stimulation. Thus, Δp_{skin} should be treated as another model parameter like permeability. However, this is not the parameter value normally specified. Instead the "unitless" or "dimensionless" van Everdingen and Hurst skin factor, s, is the parameter typically specified. The skin factor, s, is defined by

$$s = \frac{2\pi k h \Delta p_{skin}}{q B \mu} \qquad (2.7)$$

Eq. 2.7 is useful to have in the form of Eq. 2.8, which uses field units.

$$\Delta p_{skin} = \frac{141.2 q B \mu}{kh} s \qquad (2.8)$$

To illustrate the units and use of this expression, assume that $s = 5$ (remember, this is unitless), $q = 50$ STB/D, $B = 1.15$ RB/STB, and $\mu = 5$ cp. If $kh = 200$ md-ft, then

$$\Delta p_{skin} = \frac{(141.2)(50)(1.15)(5)}{200}(5) = 203 \text{ psi.}$$

If this skin pressure drop could be removed, then additional energy would become available to move hydrocarbons to the well instead of being wasted in overcoming the near-well damage.

The advantage of the skin concept is that it is a consistent measure of well condition throughout the life of a well. If a pressure transient test is performed earlier in well life and it shows a positive skin, while one performed later in life shows a less positive or negative skin, then it is correct to conclude that the well has cleaned up with time or that some intervening well work has successfully improved the well condition. Similarly, it is justified to say that changing the skin on a well from positive to negative will result in greater productivity.

The semilog approximation to the full exponential integral Theis solution can be manipulated to yield an equation for estimating skin factor once the semilog slope, m, has been established from the semilog plot of the data. The resulting expression in field units is

$$s = 1.151 \left\{ \frac{\left[p_i - p_{wf}(t = 1 \text{ hour}) \right]}{m} - \log\left(\frac{k}{\phi \mu c_t r_w^2} \right) + 3.23 \right\} \qquad (2.9)$$

It is important to keep in mind that the flowing bottomhole pressures appearing in Eq. 2.9 are measured data that include the impact of the skin factor. The notation $p_{wf(t=1 \text{ hour})}$ represents the flowing pressure obtained by extrapolating the radial flow (semilog) period straight line to a time of 1 hour. In Fig. 2.3, this would give a pressure slightly larger than the actual measured value at $t = 1$ hour. The complete set of radial flow regime analysis steps now becomes

1. Plot the measured flowing bottomhole pressure differences $[p_i - p_{wf}(t)]$ vs. time, t, on semilog paper.
2. Construct a straight line through the radial flow regime data, and calculate the slope.
3. Calculate the estimate of reservoir permeability, k, from Eq. 2.5.
4. Use the semilog slope, m, and the estimated permeability, k, in Eq. 2.9 to estimate the skin factor for the well. Note that the estimated permeability value is required because Eq. 2.9 contains k explicitly.

Values of s associated with the damage mechanisms listed above are usually less than 20. However, it is possible to see positive skins above 100 if the permeability is high enough. In the case of stimulation, the mechanisms listed above, with the exception of hydraulic fracturing, can be expected to produce values no smaller than –3. A skin factor smaller than approximately –3 can be obtained on a vertical well only through hydraulic fracturing.

Skin itself, however, should not be used as a strict indicator for well intervention. It is quite possible that spending money to remove a skin of 100 from a well is not a good decision. For high enough permeability, a 100 skin may translate into only a few pounds per square inch of actual skin pressure drop. In most real cases where this occurs, the actual bottleneck to improved well productivity is not in the wellbore connectivity to the reservoir but in the production system downstream of this interface (i.e., the string of completion equipment that starts to generate pressure loss as the fluid moves up the well to the surface). The pressure drop caused by bottlenecks in this production string ends up lumped into the well skin factor because the pressure measurement system is often located far from the actual producing interval in the well. Unless this pressure drop is corrected out of the gauge data, the true well skin factor is not known from the direct analysis of the data measured at the gauge location. This is always an issue, but it really is important only if the pressure drop between the producing interval and the gauge location is dominated by large friction pressure losses.

2.4.2 Wellbore Storage. The complete theory for predicting the influence of finite well volume on the flowing pressure response is complex and requires a great deal of mathematical machinery beyond the scope of this brief introduction (see Agarwal et al. 1970). We can, however, develop the equation for the flow regime added to our analysis tool kit by the complete theory in a simple way.

Section Takeaway

- Wellbore storage is the physical impact of finite well volume on a measured PTA response.
- Wellbore storage arises from changing well volume pressure (compressive storage) and from changing levels of fluid.
- The wellbore storage coefficient, C, can be estimated from this flow regime.
- Comparing PTA-derived C with C calculated from wellbore diagrams can help identify completion integrity issues.

Consider the well/reservoir model of Fig. 2.2 immediately prior to starting production. Because we have assumed that this well will flow once it is opened up, the wellbore must be filled with reservoir fluid all the way to the surface, and the pressure at the surface must be large enough to push fluid out of the well when it is opened up. Thus, the well can be envisioned as a tank with known volume filled with fluid under pressure. If we focus our attention on predicting the flowing bottomhole pressure for a short time immediately after the well is opened, it is reasonable to assume that the expansion of the wellbore fluid alone will dominate the amount of fluid produced and the values of the flowing bottomhole pressure. In this picture, the volume of fluid in the wellbore effectively isolates the reservoir from the influence of the imposed changes in producing rate at the surface. Assuming that the expansion of the wellbore fluid alone controls the pressure response, we can relate the flow rate to the change in bottomhole pressure using the definition of compressibility. We arrive at the simple relationship

$$\left[p_i - p_{wf}(t) \right] = \frac{B}{C} qt. \qquad \qquad (2.10)$$

We directly obtain a simple linear relationship that defines the *wellbore-storage flow regime*. In this equation, the symbol C is called the wellbore storage coefficient, which has the units of bbl/psi.

Eq. 2.10 suggests an analysis procedure that allows the estimation of C from data. A Cartesian plot of $[p_i - p_{wf}(t)]$ vs. time, t, will give a straight line with slope inversely proportional to C. This analysis is illustrated in **Fig. 2.4.**

As with the skin effect, measured data confirms the early-time well behavior predicted by Eq. 2.10. Although the extent of the wellbore-storage flow regime is usually only a few minutes, it can occasionally last for as much as a few hours. In the example, the flow regime lasts for approximately 1 hour (Fig. 2.4). Intuitively, we should expect the duration of this flow regime to be dependent on the magnitude of the skin effect. By definition, skin is a measure of the connectivity between the well and the formation. Larger skins translate into lower connectivity, causing the well to act like an isolated volume longer. Conversely, negative skin reduces the length of time for this well behavior. These qualitative trends are borne out and quantified by the full analytical treatment of the problem (Agarwal et al. 1970).

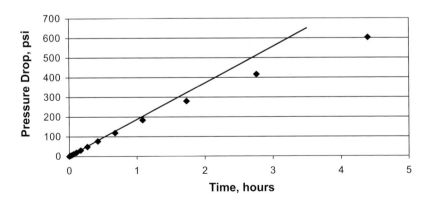

Fig. 2.4—Simulated example wellbore storage flow regime (input and estimated C = 0.025 bbl/psi).

Physically, the wellbore-storage flow regime lasts as long as the proportion of the surface rate contributed by the expansion of the wellbore fluids is essentially unity. However, the influence of the wellbore volume on the measured pressure response does not end until this same rate proportion is essentially zero. After this time, the formation is contributing the entire well rate, and the well pressure response is once again given by the Theis solution. This results in a transition period between the end of the wellbore storage flow regime and the beginning of the radial flow regime. Since the goal of most pressure transient tests is to estimate formation permeability, the existence of wellbore storage is undesirable.

This does not mean that the wellbore-storage flow regime and its accompanying analysis for wellbore-storage coefficient are of no value. Real well production systems very often include packers isolating production tubing from the rest of the well volume. If a packer leak develops, actual volume contributing to the wellbore-storage coefficient will be larger than expected. Conversely, if part of the well is filled with lower-compressibility fluid than expected (water loading in gas wells, for example), the expansive volume is effectively reduced, with a corresponding reduction in the wellbore storage coefficient.

A comparison of the wellbore-storage coefficient calculated from the wellbore-storage flow regime with that calculated from the expected well volume can be diagnostic of these problems. Unfortunately, accurately estimating the average compressibility of the expected wellbore fluid is often difficult. This makes the comparison more qualitative that quantitative. Nevertheless, it is a good exercise to complete—and if the differences are large, a cause should be tracked down.

2.4.3 Material Balance, Inflow Performance, and Boundary-Dominated Flow. All real pressure transient responses start out dictated by the volume of the well and the compressibility of the wellbore fluids. For many of the same physical reasons, the volume of the reservoir and the compressibility of the reservoir fluids control all real pressure transient responses at long time. This period is called the pseudosteady, semisteady, or *boundary-dominated flow regime*. For a single well draining a sealed finite reservoir, this regime begins when all of the reservoir boundaries begin to affect the flowing wellbore pressure. The general equation governing this flow period is (Dake 1978)

$$p_i - p_{wf}(t) = \frac{162.6qB\mu}{kh}\log\left(\frac{4A}{1.781C_A r_w^2}\right) - \frac{0.2339qBt}{Ah\phi c_t}. \quad \dots\dots\dots\dots\dots \quad (2.11)$$

In Eq. 2.11, the first term contains the drainage area of the well, A, and the Dietz shape factor (Dake 1978), C_A. The Dietz shape factor is the classic approach to accounting for drainage area shape and well position in the drainage. This approach has been largely replaced in modern analysis by methods based on derivatives.

The second term in Eq. 2.11 contains the pore volume of the reservoir, $Ah\phi$. Thus, this relationship implies that a Cartesian plot of pressure drop, $p_i - p_{wf}(t)$ vs. flowing time, t, will

---

Section Takeaway

- The boundary-dominated flow regime occurs when all reservoir boundaries are impacting the well pressure.
- Analysis of data in this flow regime yields an estimate of the pore volume connected to the well.

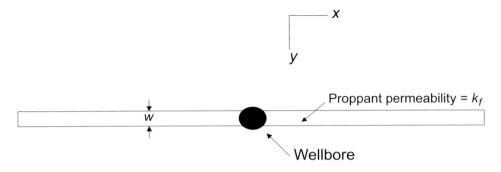

Formation permeability = k

Fracture and formation extend to infinity in x and y directions

Fig. 2.5—Conceptual model of a vertically fractured well (plane view).

plot as a straight line with a slope that is inversely proportional to the reservoir pore-volume effective-compressibility product, $Ah\phi c_{eff}$. Once the effective compressibility is estimated (Dake 1978), the pore volume of the reservoir can be calculated. We do not have immediate use for Eq. 2.11, but will find it valuable during the discussion in Chapter 3.

2.4.4 Flow Regimes for a Vertical Fracture Intercepting a Vertical Well. To derive the final flow regime relationships needed to interpret pressure transient response from a hydraulically fractured well, we consider the conceptual model shown in **Fig. 2.5.** The complete solution to the mathematical problem arising even from this simple conceptual model is well beyond what we can cover here. Instead, we will look at two further approximations (extremes) of this model and derive flow regime solutions for these extremes that will then be tested against the full solution of the complete problem (Cinco-Ley et al. 1978).

For the first approximation model, we assume that

- Fluid must flow from the formation into the fracture and then to the well.
- Flow in the formation is compressible, while flow in the fracture is incompressible. This assumption makes the problem tractable for us.
- Both fracture and formation extend to infinity.

Section Takeaway
- A vertically fractured well response can exhibit two additional flow regimes.
- The bilinear flow regime occurs before pressure drop occurs at the fracture tip.
- The linear flow regime occurs after pressure drop occurs at the fracture tip.
- Analysis of linear-flow data gives a value of fracture half-length.
- Analysis of bilinear-flow data allows an estimate of fracture conductivity.
- An estimate of permeability is required to carry out linear- and bilinear-flow regime analysis.

Under these assumptions, the pressure drop at the wellbore is given by

$$p_i - p_{wf}(t) = \frac{44.05qB\mu}{h\sqrt{k_f w}} \left(\frac{t}{\phi\mu c_t k} \right)^{1/4} \quad \ldots\ldots\ldots\ldots\ldots\ldots\ldots\ldots\ldots\ldots\ldots \quad (2.12)$$

This is the expression for the *bilinear flow regime*. The expression in Eq. 2.12 shows that a Cartesian plot of change in bottomhole flowing pressure vs. $t^{1/4}$ should give a straight line with a slope that is inversely proportional to $(k_f w)^{1/2}$. The product of fracture proppant permeability, k_f, with fracture width, w, defines the *fracture conductivity*. As we shall see when we consider the full spectrum of fractured well responses in the next section, this flow regime dominates the well response only for low values of fracture conductivity and, in practice, is usually quite short. It is also useful to note that the length of the fracture plays no role in this time period because we are considering times that are small enough for the effect of the fracture tip to be negligibly small.

The bilinear flow regime is the first of the approximations we are seeking to obtain. Since this extreme occurs for lower-conductivity fractures, it is natural to assume that the other will cover the opposite extreme: high-conductivity fractures.

If we focus our attention on fractures with high values of fracture conductivity, then a second flow regime emerges as dominant. In this case, because the fracture conductivity is assumed to be high, the pressure drop in the fracture becomes negligible. In this situation, the flowing pressure measured at the well is taken on throughout the fracture. Retaining the assumption of compressible flow in the formation, the pressure drop at the wellbore is now given by

$$p_i - p_{wf}(t) = \frac{4.06qB\mu}{h \cdot x_f} \cdot \left(\frac{t}{\phi\mu c_t k} \right)^{1/2} \quad \ldots\ldots\ldots\ldots\ldots\ldots\ldots\ldots\ldots\ldots\ldots \quad (2.13)$$

This is the equation for the *linear flow regime*.

The linear flow regime dominates the response of a fractured well with high fracture conductivity. Because fracture conductivity is assumed to be high, the impact of the tips of the fracture are essentially seen instantaneously in the well response, and the length of the fracture through the variable x_f, the fracture half-length, now influences the pressure response. This expression implies that a Cartesian plot of the change in flowing bottomhole pressure vs. the square root of time will have a slope inversely proportional to the fracture half-length, x_f.

A good picture of the pattern of flow during these flow periods is shown in **Fig. 2.6.** The first flow period pictured, fracture linear flow, is one we have not considered. Since it occurs at very early time, it has no real practical value. This flow period occurs when fluid is being produced from the fracture only and no fluid is entering the fracture from the formation. This flow regime is controlled by the expansion of the fluid in the fracture only and, hence, its length, like the length of the wellbore storage flow regime, is determined by fracture-formation connectivity and fracture volume.

The next two flow pattern pictures are for the bilinear and formation linear flow regimes (i.e., the two fracture flow regimes for which we have just derived expressions). The radial flow regime is governed by the semilog radial flow equation derived for the Theis model. We will discuss this flow regime in greater depth later. The physical picture presented in this figure is a good one to keep in mind throughout the rest of this section. These diagrams

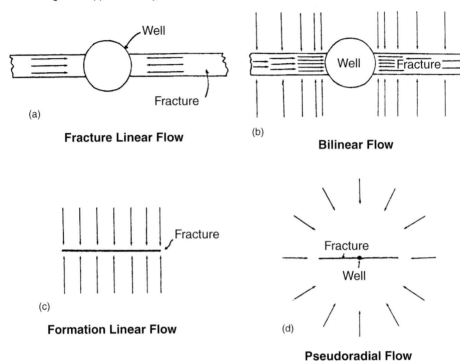

(a) **Fracture Linear Flow**

(b) **Bilinear Flow**

(c) **Formation Linear Flow**

(d) **Pseudoradial Flow**

Fig. 2.6—Flow patterns for vertically fractured well flow regimes [Fig. 3 from Cinco-Ley and Samaniego-V. (1981)].

show the dominant direction for fluid movement during the flow periods governed by the simplified relationships we have derived. We now need some method of confidently recognizing these flow regimes in measured data.

2.4.5 The Derivative and the Interpretation Algorithm. The wellbore-storage, radial, and boundary-dominated flow regimes form the basis of classic pressure transient analysis methods. Supplementing these with the bilinear and linear flow regimes extends these classic analysis methods to wells intercepted by vertical fractures with finite values of fracture conductivity, $k_f w$. However, there are practical issues with the use of these equations. By definition, these equations are approximate solutions expected to be valid for only a subset of the testing time. Though we have stated some ideas on when these approximations are valid, we have not set forward methods that allow the identification of the presence and the duration of specific flow regimes in a given measured response. This is a form of the uniqueness problem that is common in the solution of inverse problems. In modern PTA methods, this is addressed by the introduction of the logarithmic time derivative of the bottomhole pressure (Bourdet et al.1983).

Consider the five flow-regime relationships we have presented:

$$\Delta p_{wf}(t) = \frac{162.6qB\mu}{kh}\left[\log(t)+\log\frac{k}{\phi\mu c_t r_w^2}-3.23+s\right] \text{ (radial)} \quad \ldots\ldots\ldots\ldots (2.14)$$

Section Takeaway

- Practical identification of flow regime presence and length requires the derivative.
- Rules for flow regime identification exist and work.
- Systematic interpretation of PTA responses is based on an iterative algorithm.
- The diagnostic plot forms the starting and often the ending point for a PTA interpretation.
- Two new kinds of skin can occur on fractured wells.
- Improved parameter estimates require history matching.

$$\Delta p_{wf}(t) = \frac{B}{C} qt \quad \text{(wellbore storage)} \quad \dots \dots \dots \dots \quad (2.15)$$

$$\Delta p_{wf}(t) = \frac{5.615qB}{A\phi hc_{eff}} t + \frac{qBu}{2\pi kh\alpha_3} \ln\left(\frac{r_{avg}}{r_w}\right) \quad \text{(boundary-dominated)} \dots \dots \dots \dots \quad (2.16)$$

$$p_i - p_{wf}(t) = \frac{44.05qB\mu}{h\sqrt{k_f w}} \cdot \left(\frac{t}{\phi\mu c_t k}\right)^{1/4} \quad \text{(bilinear)} \quad \dots \dots \dots \dots \dots \quad (2.17)$$

$$p_i - p_{wf}(t) = \frac{4.06qB\mu}{h \cdot x_f} \cdot \left(\frac{t}{\phi\mu c_t k}\right)^{1/2} \quad \text{(linear)} \quad \dots \dots \dots \dots \dots \quad (2.18)$$

If we differentiate these equations with respect to time, t, and then multiply the resulting derivatives by t, we obtain the following for the radial, wellbore-storage, boundary-dominated, bilinear, and linear flow regimes derivatives, respectively.

$$\frac{d\Delta p_{wf}}{d\ln t} = \frac{141.2qB\mu}{kh} = \text{a constant} \quad \dots \dots \dots \dots \dots \dots \dots \quad (2.19)$$

$$\frac{d\Delta p_{wf}}{d\ln t} = \frac{B}{C} qt = \Delta p_{wf}(t) \quad \dots \dots \dots \dots \dots \dots \dots \dots \dots \quad (2.20)$$

$$\frac{d\Delta p_{wf}}{d\ln t} = \frac{5.615qB}{A\phi hc_{eff}} t \quad \dots \dots \dots \dots \dots \dots \dots \dots \dots \dots \quad (2.21)$$

$$\frac{d\Delta p_{wf}}{d\ln t} = \frac{1}{4}\left[\frac{44.05qB\mu}{h \cdot \sqrt{k_f w}} \cdot \left(\frac{1}{\phi\mu c_t k}\right)^{1/4}\right] \cdot t^{1/4} = \frac{\Delta p_{wf}(t)}{4} \quad \dots \dots \dots \dots \quad (2.22)$$

$$\frac{d\Delta p_{wf}}{d\ln t} = \frac{1}{2}\left[\frac{4.06qB\mu}{h \cdot x_f} \cdot \left(\frac{1}{\phi\mu c_t k}\right)^{1/2}\right] \cdot t^{1/2} = \frac{\Delta p_{wf}(t)}{2} \quad \dots \dots \dots \dots \quad (2.23)$$

In all of the above we have used

$$t\frac{d\,\Delta p_{wf}}{dt} = \frac{d\,\Delta p_{wf}}{d\ln t} \quad\dots\dots\dots\dots\dots\dots\dots\dots\dots\dots\dots\dots\dots\dots\dots\dots\dots\dots\dots\dots \quad (2.24)$$

to introduce the logarithmic derivative of the pressure into the expressions. From these equations, we will construct the most common set of diagnostic principles. These principles are based on plotting the change in flowing pressure and its logarithmic derivative on a log-log plot. This is usually referred to as a *diagnostic plot*. It is the first step in the modern interpretation approach to pressure transient data.

Fig. 2.7 shows a spectrum of fractured well drawdown responses. Each response is displayed on a diagnostic plot as defined previously (i.e., the derivative displayed is the logarithmic derivative). The responses were computed using a general semianalytical solution for the complete drawdown response of a vertically fractured well. This solution is well beyond the mathematical scope of this review, but the interested reader can follow the development in Cinco-Ley et al. (1978). In addition to the ability to compute the complete drawdown response, this solution also identifies an important parameter not identified clearly in any of the flow regime equations derived here. This is the dimensionless conductivity, F_{CD}, defined by

$$F_{CD} = \frac{k_f w}{k x_f}. \quad\dots\dots\dots\dots\dots\dots\dots\dots\dots\dots\dots\dots\dots\dots\dots\dots\dots\dots\dots\dots\dots \quad (2.25)$$

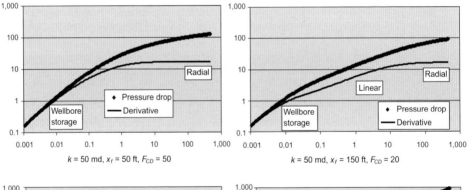

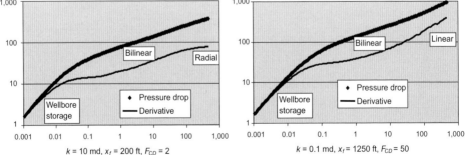

Note: Δp on x-axis; y-axis is time in hours for all plots shown.

Fig. 2.7—Diagnostic plots: typical fracture well drawdown responses.

The physical significance of this parameter is that it represents the ratio of the fracture's capacity for fluid delivery into the well to the formation's capacity to deliver fluid into the fracture. The magnitude of F_{CD} determines the presence (or absence) and the temporal extent of the linear and bilinear flow regimes. The dimensionless fracture conductivity defined by Eq. 2.25 can be introduced into Eq. 2.17 by multiplying the right side of Eq. 2.17 by $\dfrac{\sqrt{(kx_f)}}{\sqrt{(kx_f)}}$ and simplifying. The result is

$$p_i - p_{wf} = \frac{44.05qB\mu}{h \cdot \sqrt{kx_f} \cdot \sqrt{F_{CD}}} \cdot \left(\frac{t}{\phi \mu c_t k} \right)^{1/4} \quad \dots \dots \dots \dots \dots \dots \dots \dots \dots \quad (2.26)$$

Returning to Fig. 2.7, the plot in the upper left of the figure shows the response for a short, high-conductivity fracture in a high-permeability reservoir. Eq. 2.20 shows that the drawdown pressure drop and its logarithmic derivative are identical during the wellbore-storage-dominated flow regime. Each of these functions follows the same equation of the form $x = at$. On a log-log plot, these functions will overlap and form a line with unit slope. The rigorously computed response confirms the accuracy of these conclusions derived from the simplistic wellbore storage model used above. The portion of the response in each of the plots shown in Fig. 2.7 controlled by wellbore storage is readily identified and is labeled.

One other flow regime is evident in this plot, the radial flow regime. From Eq. 2.19, we see that the logarithmic derivative is expected to take on a constant value during this flow regime. Thus, on a log-log diagnostic plot, the portion of the response dominated by radial flow should exhibit a constant logarithmic derivative. The part of the response in each of the plots in Fig. 2.7 dominated by radial flow is again evident and labeled. Note that the response shown in the plot on the bottom right does not exhibit this flow regime even after 500 hours of production.

The response shown in the plot on the upper right in Fig. 2.7 exhibits a linear flow regime. From Eq. 2.23, we conclude that the value of the logarithmic derivative during this flow regime should be one-half the value of the drawdown pressure drop. In addition, the linear flow regime approximate solution dictates that both the drawdown pressure drop and its logarithmic derivative be proportional to the square root of time. Thus, these functions should both follow straight lines with ½-slope on a log-log diagnostic plot. The drawdown pressure drop and its logarithmic derivative exhibit the behaviors reflecting these diagnostic rules on the plot for $0.1 < t < 10$.

The bilinear flow regime is illustrated by the response shown in the bottom left of Fig. 2.7. Eq. 2.22 says that the logarithmic derivative should be one-quarter of the drawdown pressure drop. The bilinear flow approximate solution shows that both of these functions are proportional to the fourth root of time and, hence, they should follow lines with ¼-slope on a log-log diagnostic plot. These functions exhibit the behavior predicted by the approximate solution for $0.3 < t < 5$.

The comments in the last four paragraphs form the basic diagnostic rules for identifying the presence of the four vertically fractured well flow regimes. These rules are summarized in **Table 2.1** along with comments concerning correct analysis and parameters estimated. The temporal order of the flow regimes accurately reflects the physical assumptions used to arrive at these approximations. For example, the bilinear flow regime occurs prior to the formation linear flow regime because the bilinear flow regime's governing equation arises

TABLE 2.1—FLOW-REGIME DIAGNOSTICS SUMMARY				
Flow Regime	Rules for ID	Analysis Method	Parameter Estimated	Comments
Wellbore storage (WBS)	p and log-derivative overlay on unit slope line	Δp vs. t on Cartesian plot	Cartesian plot of C	Occurs at early time, always before other flow regimes
Bilinear (BL)	Δp and log-derivative separated by ¼ and follow lines with ¼ slope	Δp vs. $t^{1/4}$ on Cartesian plot	$k_f w$, product of proppant permeability and fracture width	Occurs after WBS and before linear flow; calculating $k_f w$ requires estimate of k and x_f
Linear (LF)	Δp and log-derivative separated by ½ and follow lines with ½ slope	Δp vs. $t^{1/2}$ on Cartesian plot	x_f, fracture half-length	Occurs after BL and before RF; calculating x_f requires estimate of k
Radial (RF)	Log-derivative constant	Semilog plot of Δp vs. t	k, formation permeability	Last transient flow regime; requires estimate of formation thickness, h

when the tip of the fracture has not begun to influence the wellbore pressure response. The fact that the complete solution exhibits periods that follow the behavior predicted by the flow regime equations is taken as a justification for the applicability of these assumptions for some portion of the expected drawdown response.

The use of the diagnostic plot combined with the flow period identification rules essentially eliminates ambiguity for most cases (modifications required for the most common exceptions are simple and will be discussed later). The real problem that arises with the use of these rules for data analysis is not lack of applicability but, rather, nonuniqueness arising from the fact that most measured responses do not exhibit all four of the flow periods required to determine model parameters. This is readily observed in Fig. 2.7, where none of the four responses shown contains all four of the vertically-fractured-well flow regimes.

It should also be emphasized that in practical interpretation and analysis, the rules that relate to the derivative are used much more than the rules for the pressure drop or the rules for the relationship between the pressure drop and its derivative. Emphasis on the derivative rules occurs because the pressure drop can be distorted away from exhibiting a half- or quarter-slope by the presence of poor communication between the fracture and the wellbore perforations or between the fracture and the formation. These reasons for poor communication form other sources for skin factors. In the case of poor fracture-perforations communication, it is called *plug skin,* and in the case of poor fracture-formation communication, it is called *fracture face skin.* Proppant overdisplacement and subsequent closing of the fracture near the well can cause the former. The latter is usually caused by interactions between the fracturing fluid and the fluids in the formation. We will further consider plug and fracture face skin later in one of the example interpretation exercises.

Consider again the response in the upper left of Fig. 2.7. This response represents a short, high-conductivity fracture in a moderate-permeability formation. However, the response contains neither the bilinear nor the linear flow regimes that specifically reflect vertically-fractured-well behavior. Could this response be equally well explained by a simple vertical well model [i.e., with the complete solution for a well impacted by wellbore storage and skin presented by Agarwal et al. (1970)], or does it require a vertical fracture well model? What range of fracture half-length and conductivity are consistent with the response (i.e., honor the wellbore storage period, radial flow period, and the values of C and k they imply)? We will consider the second question first.

Fig. 2.8 is a diagnostic plot [i.e., a log-log plot of pressure drop and its logarithmic derivative vs. time, comparing the original data (black) from the upper right plot in Fig. 2.7 with two other responses also calculated from the complete vertically fractured well solution of Cinco-Ley et al. (1978)]. The response shown in green (dots are the pressure drop; the solid line, the logarithmic derivative) is for a fracture with $x_f = 40$ ft and $F_{CD} = 60$, while the response shown in red is for a fracture with $x_f = 60$ ft and $F_{CD} = 40$. To honor the wellbore-storage and radial flow regimes present in the original response, the permeability and the wellbore-storage coefficient have not been changed. The model parameter variations were chosen so that the two sensitivities would produce approximately the same pressure drop as the original response at later time. Based on physical intuition, justified by the results shown, it should be clear that this requires opposing changes in the values of x_f and F_{CD}.

Fig. 2.8 indicates that in comparison to the exact original response, the 20% parameter variations are recognizable, especially if greater weight is given to the differences in the derivatives during the transition period between the wellbore storage and the radial flow regime. However, as we will see later, measured pressure transient responses are not generally so well defined, particularly with respect to values of the derivative. Nevertheless, the reasonable sensitivity demonstrated in Fig. 2.8 leads to the idea that even if flow periods are missing from a given response, the parameters that should be determined from these flow periods can still be estimated by trial-and-error matching of the given response. This trial-and-error matching process is called *history matching* and is widely used. In history matching, model parameters are adjusted by a rule- and experience-guided trial-and-error process until the computed response visually matches the measured test response. Matching is usually done by comparing computed and measured responses on a log-log diagnostic plot at a minimum. Response comparisons on typical flow regime plots (semilog, for example)

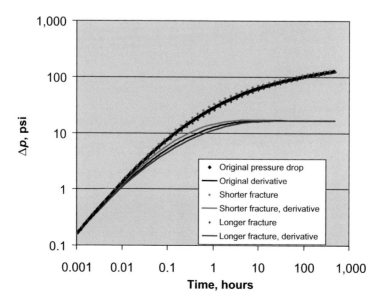

Fig. 2.8—Sensitivity of diagnostic plot to changes in F_{CD} and x_f.

and subsequent further trial-and-error adjustments to model parameters are also often made. The accuracy of the history-matching approach obviously depends on the quality of the measured data and, perhaps not so obviously, on which flow periods are present, the clarity of their definition, and the number of missing flow periods. For example, if the radial flow period in the original response were missing, as it is in the response shown at the lower right of Fig. 2.7, then no flow period estimate for the permeability value would be available. This makes the history match much less determinate. It is left to the reader to imagine the added impact on the matching process if the derivative exhibited a +/– 20% band of variation in its values.

We return now to consider the first question asked above. **Fig. 2.9** shows a match of the fractured-well response we have been working with using Agarwal's complete solution to the radial flow with wellbore-storage and skin model. Achieving this match required the use of the wellbore-storage coefficient and permeability values determined by the appropriate flow periods. In addition, the skin value determined from the radial flow period was also input. The skin value calculated is –4.3 using Eq. 2.9. Thus, the answer to our question is most certainly yes. The response can be explained by a simpler model. We also see that the radial-flow regime equation including skin (Eq. 2.4) can be used to predict not only the derivative but also the pressure drop for a vertically fractured well, once this flow period develops. Finally, it follows from this last statement that once the fractured well response reaches the radial flow period, the fracture itself is equivalent to a negative skin.

Though it was possible to predict the entire response of the specific fractured well case that we have been examining with a model solution derived for a nonfractured vertical well, this generally turns out to be correct only for the part of the response that occurs after the onset of radial flow. If the transition zone between the wellbore storage and radial flow regimes in Fig. 2.9 had been dominated by one or more of the vertical-fracture flow regimes

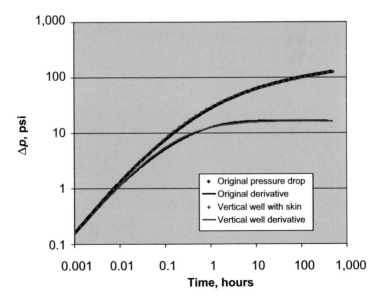

Fig. 2.9—Match of fracture well response using the Agarwal storage and skin model.

(bilinear or linear flow), then the match of the data with the simpler nonfractured well model would not have been so satisfactory.

As was alluded to above, the fracture acts as an equivalent skin once the well response reaches the radial flow regime and beyond. The value of this negative skin can be predicted if an estimate of fracture length, x_f, fracture conductivity, $k_f w$, and permeability, k, are known. If Eq. 2.7 is solved for skin pressure drop and is added to the radial flow regime equation (Eq. 2.4), we obtain

$$p_i - p_{wf}(t,s) = -\frac{qB\mu}{4\pi kh}\left[\ln\left(\frac{r_w^2}{4\eta t}\right)+\gamma\right]+\frac{qB\mu}{2\pi kh}\,s.$$

This equation can be rearranged with simple algebra to yield

$$p_i - p_{wf}(t,s) = \frac{qB\mu}{4\pi kh}\left\{\ln\left[\frac{4\eta t}{\left(r_w'\right)^2}\right]-\gamma\right\} \quad \dots\dots\dots\dots\dots\dots \text{(2.27)}$$

In Eq. 2.27, r_w' equals $r_w e^{-s}$ and is defined as the *equivalent wellbore radius*. This is the value needed in the nonfractured well radial-flow regime equation to predict the pressure response of a fractured well after the onset of radial flow. Since the skin associated with a fractured well is negative, the equivalent wellbore radius of a fractured well is larger than the actual drilled well radius by the factor e^{-s}. Cinco-Ley et al. (1978) presented a correlation between F_{CD} and r_w'/x_f which is reproduced here as **Fig. 2.10**. In this figure, first note that the x-axis is the equivalent of the dimensionless fracture conductivity, F_{CD}, and second, note that for $F_{CD} > 30$, we have the following simple relationship for equivalent wellbore radius.

$$r_w' = \frac{x_f}{2} \quad \dots\dots\dots\dots\dots\dots\dots\dots\dots\dots\dots \text{(2.28)}$$

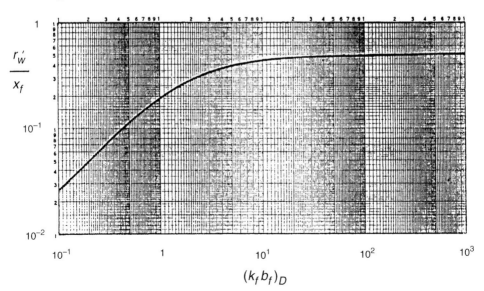

Fig. 2.10—r_w'/x_f vs. F_{CD} [after Cinco-Ley et al. (1978)].

The equivalent wellbore radius given in Eq. 2.28, valid for the high range of F_{CD} values, is called the *infinite-conductivity equivalent wellbore radius* because it is independent of the magnitude of F_{CD} in this range. For consistency, fractures with dimensionless conductivity above 30 are referred to as infinite-conductivity fractures.

Because Eq. 2.27 was derived independent of the source of the skin factor and whether the skin factor was negative or positive, the equivalent wellbore radius definition can account for all types and magnitudes of skin factor. As was mentioned, if the skin factor is negative, then the equivalent wellbore radius is larger than the actual wellbore radius. Accordingly, if the skin is positive, then the equivalent wellbore radius is smaller than the actual drilled radius. Thus, the effect of skin factor on pressure drop is seen to be equivalent to physically drilling a larger or small borehole.

The equivalent (or effective) wellbore radius will be put to further use later during the discussion of rate-time decline analysis (RTDA) in Chapter 3.

Before moving on, it will be of value to collect the steps for the analysis of a pressure response, woven somewhat implicitly throughout the above discussion, into a clear algorithmic approach. These steps are as follows:

1. Make a log-log diagnostic plot of the measured well pressure response.
2. List all appropriate models based on the shape of the derivative response and choose a model to test.
3. Locate the flow regimes present using the rules delineated in Table 2.1.
4. Make the flow regime plots as described, and calculate the appropriate parameter from straightline analysis.
5. History match the total pressure response to obtain the best estimate for all parameters, including those determined by flow regime analysis.
6. Examine best-match parameters in context.

These six steps form the *interpretation and analysis algorithm* for all pressure transient data. Step 2 involves developing a list of possible models that should be tested to explain the response. For the vertically-fractured-well case, we would normally assume that the best model to start with would be the vertically-fractured-well model. Step 6 involves a critical evaluation of the match quality and the values of the match parameters in the context of all the information known about the well-reservoir system that the measured response represents. The goal of Step 6 is to decide whether another of the possible models identified in Step 2 should be carried through the quantitative steps of the algorithm, or the last model tested should be accepted as the "best" model for the data. We have no need for these steps of the algorithm here because we are always going to be working with vertically fractured wells and we will only be using one vertically-fractured-well model to interpret responses. However, a brief discussion of available vertically-fractured-well models is justified at this point.

Most pressure transient interpretation software incorporates three vertical-fracture-well models, the *uniform-flux, infinite-conductivity, and finite-conductivity models*. The uniform-flux model assumes that the rate at each point along the fracture is the same and is given by the total well rate divided by the fracture length. The infinite-conductivity model is obtained from the uniform-flux model by evaluating the pressure drop at a particular point along the fracture. Neither of these models contains F_{CD} as a parameter; it is not in the uniform-flux model because there is no real flow down the fracture in this model, and it is not in the infinite-conductivity model because it is fixed at the value infinity. For clarity, the model we have been working with throughout is the finite-conductivity model. The reader who has the interest should consult Raghavan (1993) for the details of the uniform-flux and infinite- conductivity models. For our purposes, a simple comparison of the three model responses will be sufficient. **Fig. 2.11** shows this comparison for the model parameters used to generate the upper left plot in Fig. 2.7. The differences between the finite-conductivity model response and the infinite-conductivity model response for these parameters are negligible. In general, the only reason to use the infinite-conductivity model to interpret vertically-fractured-well data is the simplicity of having one fewer model parameters. This is often outweighed by the need to have some estimate of the real level of in-situ fracture conductivity. We pay a premium in proppant quality and/or proppant quantity to generate high levels of conductivity in situ. Hence, it is reasonable and valuable to strive to quantitatively evaluate the level of conductivity actually achieved when a design is executed. As noted above, there is very little theoretical productivity advantage to achieving $F_{CD} > {\sim}30$. If a given design aimed at generating an infinite-conductivity fracture achieves a larger value of in-situ dimensionless conductivity, then there exists an opportunity to adjust the design to lower the cost while still meeting the goal of placing an effectively infinite-conductivity fracture.

Section Takeaway

- Analysis of fractured-well data often requires an externally derived estimate of permeability.
- Reasonable estimates of fracture model parameters come from flow regime analysis.
- Model-parameter estimates need to be vetted and improved by means of history matching.
- Algorithmic analysis is easy to work through and yields reliable results, at least for quality data.

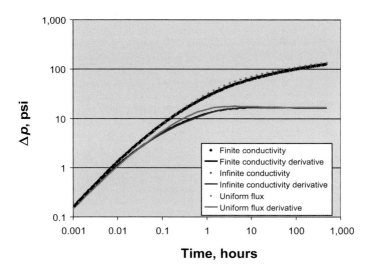

Fig. 2.11—Comparison of various fracture model responses.

There are noticeable differences between the uniform-flux model response and the other two responses. The uniform-flux model has the reputation of doing a better job of matching responses from unpropped fractures such as those created in injection wells when formation parting pressure is exceeded or those created by pumping acid at rates high enough to initiate formation fracturing (i.e., pumping above "matrix rates"). These types of fracture stimulations, though not rare, are not the focus here, and further discussion is not warranted. Instead, we now turn to an example interpretation to help generate some experience with the interpretation and analysis process and to gain some understanding of the limitations.

2.4.6 Example: Drawdown Test Interpretation. The data set chosen for this example is the same as that shown in the lower right of Fig. 2.7. This data set contains wellbore-storage and bilinear and linear flow regimes, but it does not contain a radial flow regime. To illustrate the confirmation of these flow regimes, **Fig. 2.12** shows a log-log diagnostic plot of this data set (pressure drop in red and logarithmic derivative in blue) with a line of unit slope struck in the early wellbore-storage period, a line of quarter-slope struck in the intermediate time period, and a line of half-slope struck through the late-time data. In the interpretation program we are using here, the regions of the various appropriate slopes in the derivative are found by sliding a line of the desired fixed slope across the diagnostic plot until a region in the derivative data is found to fit that slope. In doing this, we need to keep in mind the temporal relationships between the flow regimes expected [i.e., wellbore storage before quarter-slope, quarter-slope before half-slope, and half-slope before the zero derivative slope (constant derivative value) representing radial flow].

Recalling the summary in Table 2.1, we note that to analyze the half-slope period for an estimate of x_f an estimate of permeability is required. Similarly, to analyze the quarter-slope period for a value of F_{CD}, we need the resulting estimate of x_f in addition to an

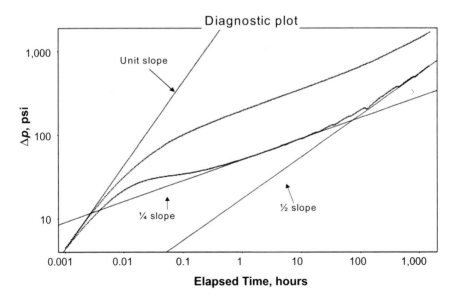

Fig. 2.12—Diagnostic plot: drawdown example.

estimate of permeability. Because this response does not contain a radial-flow period to give an estimate of permeability, another source for a value of this model parameter must be found. Typical possible sources for this estimate are a prefrac test of some form, a value from a test on a nearby well in the same formation, log- or core-derived values appropriately corrected for net-overburden and water saturation effects, or an expected value for the formation in the area. From some source, we will assume that a permeability of 0.1 md has been estimated, and we will proceed with the analysis.

Figs. 2.13 through 2.15 show the straightline analyses appropriate for the wellbore-storage, linear (half-slope), and bilinear (quarter-slope) flow regimes, respectively. The

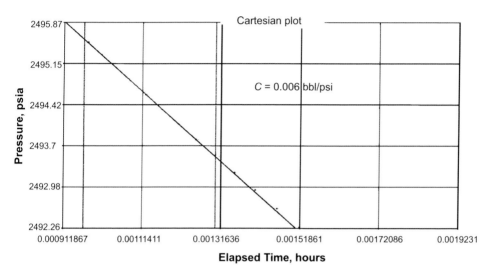

Fig. 2.13—Wellbore storage analysis: drawdown example.

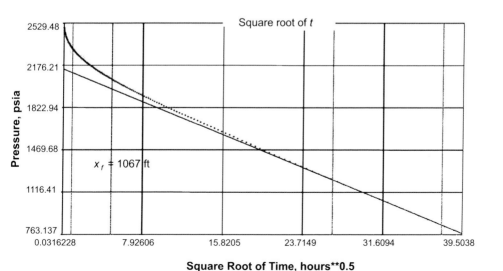

Fig. 2.14—Linear flow regime analysis: drawdown example.

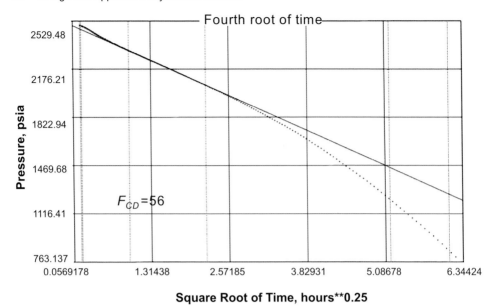

Fig. 2.15—Bilinear flow regime analysis: drawdown example.

equations used to do the flow regime analyses are Eqs. 2.15, 2.18, and 2.17, listed in the same order. In constructing these plots, it is obviously important to draw the straight lines through the parts of the data set delineated as correct on the log-log diagnostic plot. These analyses give $C = 0.006$ bbl/psi, $x_f = 1067$ ft, and $F_{CD} = 56$.

A comparison between the data set and the response predicted using the parameter values derived from flow regime analysis with $k = 0.1$ md is shown in **Fig. 2.16.** It is clear that the parameter values obtained from the straight-line analyses need to be improved. We do this by adjusting the values of C, F_{CD}, and x_f until the predicted response overlays the measured data set (i.e., by history matching). When doing this by iterative trial and error (rather than by some form of mathematical regression), the following rules are usually the most successful.

1. First, adjust the permeability to get the best match of the radial flow regime if present. Remember, permeability is inversely proportional to the value of the logarithmic derivative during the radial flow regime, so if the predicted derivative is below the level of that in the data, decrease the permeability. If above, increase the model value of permeability. If no radial flow regime is present, honor the permeability assumed to obtain the parameter estimates from the other flow regimes.
2. Adjust the fracture half-length to get the best match of the identified half-slope region in the data set. Again, if the model predicts a half-slope region above (to the left of) that in the data set, increase the value of x_f. If the predicted half-slope is below (to the right of) that in the data set, decrease the fracture half-length.
3. Next, adjust the value of C so that the predicted and measured unit slope periods match. If the predicted unit slope is above (to the left of) the data set unit slope, increase C. Conversely, C is too large if the predicted unit slope lies below (to the right of) the corresponding period in the measured data.

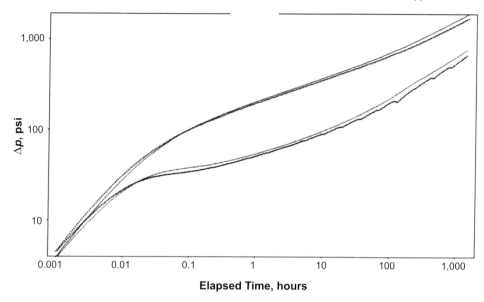

Fig. 2.16—Preliminary match using flow regime derived parameter values: drawdown example.

4. Finally, adjust the value of F_{CD} so that the quarter-slope periods match. Decrease F_{CD} to move the predicted model quarter-slope period up (to the left) and increase it to have the opposite effect.

Most interpretation programs have automated history-match options that should work well when the initial parameter estimates are reasonably good, as they are in this case. However, we have not used any such automated regression approach here; we have followed the trial-and-error history-matching rules to yield the final match and parameter estimates shown in **Fig. 2.17.** This does not imply that automated matching approaches are less preferable or should not be used. The algorithms programmed into most commercial packages are very effective if used with judgment. Nevertheless, comparing our final trial-and-error estimates with the known values reported on the lower right plot of Fig. 2.7, we conclude that our analysis has yielded very good results. Note, however, that this success was achieved by knowing a good estimate of permeability. Other matches with lower values of permeability are possible. Usually a final match of this type of data also requires consideration of other information such as the range of F_{CD} and x_f expected from the design and subsequent execution of the stimulation job.

The quality of this final match has been judged just by visual inspection of the log-log plot comparing the predicted and measured data. Though sufficient for this synthetic case, matches of real field data should also be inspected on a Cartesian plot comparing at least the values of predicted and measured pressure. Interpretation programs usually present statistics on the quality of the match or confidence intervals on the parameter estimates (Abbaszadeh and Kamal 1988). This kind of information should be examined before declaring a history match completed.

2.4.7 Pressure Buildup Testing. The theory and practice outlined so far is limited to tests fitting the drawdown paradigm (i.e., initial production at a constant rate of a slightly

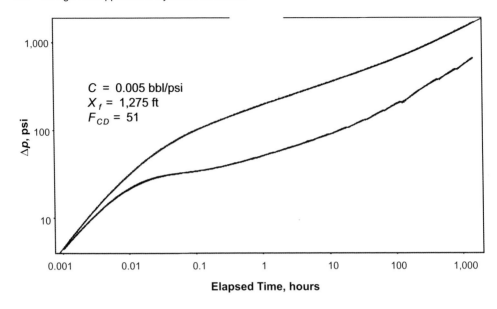

Fig. 2.17—Adjusted final match: drawdown example.

compressible fluid from a reservoir with constant properties). These requirements are serious barriers to practical application of the theory. For example, because well condition as indicated by skin factor can and often does change with time, repeat testing is a capability we would like very much to have. However, repeat testing of even a single-well reservoir development is impossible using the theory as it stands, unless we are prepared to stop producing the well until formation pressure has equilibrated, and then begin the testing sequence. What we need is some way to account for the impact of prior production on the data we would collect during some later rate-change period where we are collecting data to analyze. If we had a method to account for prior production, then we could just change the well's production rate to some new value at the desired time, collect the data during this constant rate period, and analyze the recorded data, taking into account the previous production rates.

The simplest and most popular approach to collecting data in this way is to change the rate to zero after a period of production at an essentially constant rate. This type of test is called a *single-point flow and buildup test*. **Fig. 2.18** illustrates the pressure response and

---

Section Takeaway

- Pressure buildup testing is by far the most common and reliable source of PTA data.
- The solution to the potentially multirate pressure buildup problem is constructed from single-rate drawdown solutions.
- Pressure-buildup data are analyzed by using time functions to transform the data.
- Drawdown paradigm rules and methods apply once data are transformed.
- A careful calculation and plotting approach for derivatives on diagnostic plots is required

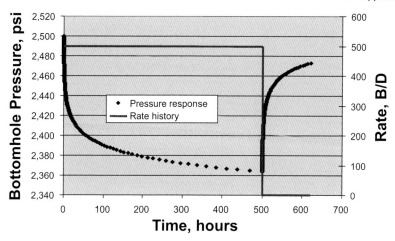

Fig. 2.18—Ideal flow and buildup test response.

rate history for the ideal single-point flow and buildup. In this example, the well is produced at 500 B/D for 500 hours, after which the rate is instantaneously reduced to zero for a subsequent 125 hours. The data collected during the 125-hour zero-rate period is called the *pressure buildup response*, and these are the data we are interested in interpreting and analyzing.

We claim that an expression for the pressure response during an entire single-point-flow and buildup test can be derived by appropriately combining ideal drawdown test solutions. In particular, we can show formally that the single-point-flow and buildup pressure drop, $\Delta p_{\text{SPFB}}\left(r_w,\, t\right) = p_i - p_{wf,\,\text{SPFB}}\left(r_w,\, t\right)$, is given by

$$\Delta p_{\text{SPFB}}\left(r_w, t\right) = \Delta p_{\text{DD}}\left(r_w, t\right) - \Delta p_{\text{DD}}\left(r_w, t - t_p\right) \quad \dots\dots\dots\dots\dots\dots\dots \quad (2.29)$$

In Eq. 2.29, $\Delta p_{\text{DD}}\left(r_w, t\right)$ is the solution derived for the drawdown and t_p is the length of time the well is produced at the rate of interest. This time value is called the *producing time*. The only difference between the two terms appearing on the right side of Eq. 2.29 is the time at which the two drawdown solutions are evaluated. Note that Eq. 2.29 implicitly uses the formal convention

$$\Delta p_{\text{DD}}\left(r_w, t\right) = 0. \text{ for } t \le 0. \quad \dots\dots\dots\dots\dots\dots\dots\dots\dots\dots\dots\dots\dots\dots \quad (2.30)$$

This convention reflects the physical reality that for $t \le 0$, the reservoir pressure is everywhere equal to the initial pressure, since production has not yet started.

Using the physically motivated convention (Eq. 2.30), we can qualitatively motivate our assertion that Eq. 2.29 yields the pressure response during a single-point flow and buildup test in two ways. First, during the time period when $t < t_p$, the second term on the right side of Eq. 2.29 is identically zero according to Eq. 2.30. Thus, we have

$$\Delta p_{\text{SPFB}}\left(r_w, t\right) = \Delta p_{\text{DD}}(t) \text{ for } t < t_p.$$

This is certainly the desired result since, for this time period, the desired pressure response is the drawdown response. Second, since each of the two drawdown terms on the right of

Eq. 2.29 represents a well producing at the same rate, the net rate resulting from the difference for times greater than the producing time, t_p, must be zero. Thus, Eq. 2.29 honors the desired rate history. In fact, these two observations are the essential elements of a formal proof of the ability of Eq. 2.29 to yield the pressure response we want. Further details can be found in Raghavan (1993), for example.

The single-point-flow and buildup problem is extremely important to pressure transient testing since this type of testing overwhelmingly dominates in practice. Experience has consistently shown that the data collected during nominally constant-rate flowing periods (drawdowns) is erratic, difficult to differentiate and, hence, difficult to interpret and analyze. These difficulties are usually due to fluctuations in producing rate caused by variations in the state of the production system, changing sales line pressure, unstable separator conditions, or operator adjustments. It is quite difficult to meet the "production at a constant rate" dictate of the drawdown paradigm under actual field conditions, and it is equally difficult in most instances to provide fully detailed, quantitative measurements of these variations in rate. This is the great advantage offered by the zero-rate part of a pressure buildup test (the pressure buildup response); holding the production rate constant at zero is quite easy. Consequently, measured pressure buildup responses tend to be smooth, more easily differentiated, and, hence, more amenable to an interpretation and analysis process based on derivatives.

These advantages for pressure buildup testing do not imply that rate variations during the flow period prior to the pressure buildup period do not impact the measured buildup response. However, experience has shown that we can interpret and analyze pressure buildup data with much less detailed knowledge of the rate variations prior to the buildup period than we would require for the interpretation and analysis of the flowing pressure data directly subjected to these rate changes. More on this issue will be illustrated through a simulated example later in this section.

The expression given in Eq. 2.29 needs some modifications to be of practical value. What we would really like to have is an expression governing the behavior of the pressure response during the buildup period only. We would also seek an expression that governs the pressure change referenced to the pressure at the start of the pressure buildup period rather than a pressure change referenced to initial pressure.

As noted above, the functions on the right of Eq. 2.29 are the pressure changes from the *same* constant-rate drawdown problem with the second solution shifted in time from the first. For convenience, we now replace $\Delta p\,(r_w, t)$ with $\Delta p_{wb}(t)$ on the left and ignore the explicit subscripts on the left to write

$$\Delta p_{wb}(t) = \left[p_i - p_{wb}(t)\right] = \left[p_i - p_{wb}(t)\right]_{DD} - \left[p_i - p_{wb}(t - t_p)\right]_{DD}.$$

In this expression, the subscript *wb* refers to wellbore pressure. Subtracting $p_i - p_{wb}(t_p)$ from both sides and rearranging gives

$$\left[p_{wb}(t) - p_{wb}(t_p)\right]_x = \left[p_i - p_{wb}(t - t_p)\right]_{DD} - \left[p_i - p_{wb}(t)\right]_{DD} + \left[p_i - p_{wb}(t_p)\right].$$

If we now focus on the time period after t_p by setting $\Delta t = t - t_p > 0$, the above becomes

$$p_{wb}(t_p + \Delta t) - p_{wb}(t_p) = \left[p_i - p_{wb}(\Delta t)\right]_{DD} - \left[p_i - p_{wb}(t_p + \Delta t)\right]_{DD} + \left[p_i - p_{wb}(t_p)\right] \ldots (2.31)$$

This expression (Eq. 2.31) fulfills the two goals we set above. It references the pressure change during the buildup period to the pressure at the instant of shut-in, $p_{wb}(t_p)$, and it is an expression governing the pressure change during this period explicitly since the time $t_p + \Delta t$ spans this time frame only as Δt increases.

A digression on notation is appropriate at this point. The reader should note when reviewing any of the references that the following equivalent notations hold *for the left side of Eq. 2.31 only.*

$$p_{wb}(t_p + \Delta t) \equiv p_{ws}(\Delta t); p_{wb}(t_p) \equiv p_{wf,s}.$$

By definition, $p_{ws}(\Delta t)$ denotes the measured wellbore pressure during the shut-in period and $p_{wf,s}$ denotes the measured flowing wellbore pressure at the instant of shut-in.

Though we have implicitly derived Eq. 2.31 using constant-rate solutions on the right side that assume a line source well in an infinite reservoir, it is important to realize that Eq. 2.31 is actually much more general. As long as the underlying constant-rate problem is linear, Eq. 2.31 will be the correct solution for the pressure buildup response. In particular, Cinco-Ley's (1978) solution for the constant-rate drawdown response of a finite conductivity vertically fractured well can be used in Eq. 2.31 in place of the simple constant-rate line-source well solutions to obtain the equation governing the pressure buildup response of a finite-conductivity vertically fractured well.

On first consideration, Eq. 2.31 is not an obviously promising expression. To begin to understand how we can use this equation to interpret and analyze pressure buildup data, the standard approach is to assume that the constant-rate solutions on the right side can be approximated by a flow period equation. For example, if we assume the semilog equation for the radial-flow period holds for all three of the solutions on the right side of Eq. 2.31, we can show that

$$p_{wb}(t_p + \Delta t) - p_{wb}(t_p) = \frac{162.6qB\mu}{kh} \log\left(\frac{t_p\Delta t}{t_p + \Delta t}\right) \quad \dots\dots\dots\dots\dots\dots\dots \quad (2.32)$$

Eq. 2.32 allows us to make two important observations. First, for times when the constant-rate solutions on the right side show semilog or radial-flow periods, the pressure buildup response also follows a similar semilog or radial-flow equation. In fact, a semilog plot of the left side of Eq. 2.32 against the time function inside the log on the right side will yield a straight line with the same slope as was found for the constant-rate drawdown solution. Second, the derivative of the pressure buildup response with respect to $\log\left(\frac{t_p\Delta t}{t_p + \Delta t}\right)$ yields the same constant value as the log-time derivative of the constant rate drawdown response. This implies that at least one of the data interpretation rules useful for drawdown data can be extended to pressure buildup data if the more complex time function in Eq. 2.32 is used when taking the derivative.

The particular grouping of producing time, t_p, and shut-in time, Δt, appearing as the argument of the logarithmic function in Eq. 2.32 is called Agarwal's equivalent time and is denoted by t_{eq} (Agarwal 1980). Agarwal's equivalent time will be the buildup time function used throughout this discussion since it is the more natural choice to accompany the use of the pressure change referenced to the pressure at shut-in. The Horner (1951) time function is also used throughout the literature, and either function can be found as the default buildup time function of choice in commercial interpretation and analysis software packages.

Returning to Eq. 2.31, we can readily make one approximation that will move us much closer to our goal. If we assume that $t_p \gg \Delta t$ (i.e., the well is produced for a long time, at least relative to the intended length of the shut-in period, prior to starting the pressure buildup test), then the last two terms on the right side of Eq. 2.31 can be taken as equal and the equation reduces to

$$p_{ws}(\Delta t) - p_{wf,s} = \left[p_i - p_{wb}(\Delta t)\right]_{DD} \dots\dots\dots\dots\dots\dots\dots\dots\dots\dots\dots\dots \quad (2.33)$$

Note that we have now introduced the notation standard in the literature on the left side of Eq. 2.33.

This equation teaches us two things. First, if producing time is much longer than the true length of the pressure buildup test, then the pressure change during the shut-in period referenced to the final flowing pressure is essentially equal to the pressure drop predicted by the single-rate drawdown solution. Hence, all of the interpretation and analysis machinery we have developed for the single-rate drawdown can be used directly in this case. Second, independent of the length of producing time, there will always be some part of the early-time buildup pressure rise predicted precisely by the single-rate drawdown solution. For example, we can expect that if the well is produced far past the wellbore-storage-dominated period, then the pressure rise during the shut-in period would exhibit a wellbore-storage period identical to the drawdown solution wellbore-storage period. Furthermore, we would recognize and analyze this wellbore-storage period in buildup response by using the rules and equations developed earlier for the drawdown solution.

Though these conclusions derived from Eq. 2.33 are extremely satisfying, they are not quite good enough to give us a generally practical approach for buildup data interpretation and analysis. The problem is that most single-point flow and buildup tests are run so that the ultimate length of the pressure buildup is larger than the length of the initial flow period. In fact, because at least some part of the hydrocarbon stream is often vented or flared when testing newly drilled and stimulated wells, the flow-period part of a single-point flow and buildup is often quite short due to greenhouse-gas-emissions issues. What would be ideal is to find a way of plotting the pressure rise during the buildup so that the early parts of the response where we expect Eq. 2.33 to hold would merge cleanly into the later parts of the response where this expression does not hold. In addition, we would like to do this in such a way that all of the drawdown interpretation and analysis methods we have would carry over well enough for us to get reasonable model-parameter estimates using them.

More precisely, our question is: "How do we plot pressure buildup data so that the full array of drawdown interpretation and analysis methods can be applied?" Developing the answer to this question is too difficult for the discussion here. Again, the interested reader is referred to Raghavan (1993) and the references there. Here we can only summarize these results in the form of a plotting rule. The rule for pressure buildup responses following single-point flow periods is: *Calculate the logarithmic derivative of the pressure response using equivalent time and then plot both this derivative and the pressure rise,* $p_{ws}(\Delta t) - p_{wf,s}$, *against elapsed shut-in time,* Δt. Following this rule will give the best log-log diagnostic plot of the pressure buildup response in the sense that flow-period identification and analysis methods developed for single-rate drawdown data will be followed as closely as possible by the pressure buildup data.

To give some idea of how well this plotting approach works, **Fig. 2.19** presents the vertical fracture drawdown response for three different producing times and compares the

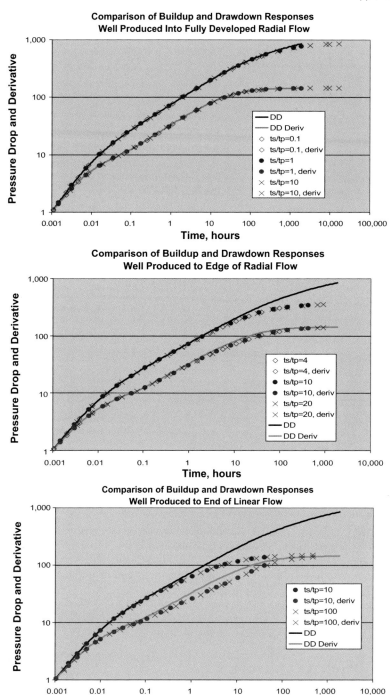

Fig. 2.19—Comparison of drawdown and buildup responses showing impact of producing and shut-in times.

subsequent buildup response for a number of shut-in times. In each of the plots shown, the solid black curve is the drawdown pressure drop, while the solid green curve is its logarithmic derivative. For easier comparison, the drawdown response shown on each of the plots is for the longest producing time. The circles, diamonds, and "x"-shapes represent the pressure buildup data computed from Eq. 2.31 using the Cinco-Ley et al. (1978) drawdown solution. These responses are parameterized by the ratio total shut-in time, t_s divided by producing time, t_p.

The top plot in the figure illustrates the comparison between the drawdown and the buildup responses after the well is produced into fully developed radial flow. This is the drawdown for the longest producing time used in the figure and is the drawdown curve shown on the other plots. The drawdown derivative indicates about 1 full log-cycle of radial flow data (constant derivative at the end of the response). The diamonds, circles, and "x"-shape data points show the impact on the comparison as total shut-in time increases. Note that even when total shut-in is 10 times longer than producing time, the pressure buildup response and its derivative remain very close to the drawdown shape and values.

The middle plot in Fig. 2.19 shows the impact on the comparison when the producing time is only long enough to reach the edge (beginning) of the radial flow period, 50 hours for this response. In this case, the buildup derivatives, calculated and plotted as described above, still follow closely the drawdown derivative even into and throughout the radial flow period. However, the buildup pressure rise begins to drop below the drawdown pressure drop as shut-in time approaches the producing time and is well below the drawdown for longer shut-in times. A brief digression back to Eq. 2.31 will allow us to explain this.

Considering the first two terms on the left-hand side of Eq. 2.31, we see that as shut-in becomes much larger than producing time, these terms become approximately equal and cancel out. This means that the pressure rise cannot be any larger than the last term in Eq. 2.31, the drawdown pressure drop calculated at the value of producing time. Hence, we should expect the buildup pressure rise to fall below the drawdown pressure as shut-in time increases and the pressure rise stabilizes at its largest possible value.

This leads us to the last plot in Fig. 2.19. Here the producing time prior to beginning the pressure buildups shown is only long enough to reach the end of the linear flow period, 5 hours for this response. For this case, the buildup derivative reproduces the drawdown derivative only at early and late times. For intermediate times, the comparison is poorer, though the shape is approximately correct.

The buildup pressure rise is a very poor approximation to the drawdown pressure drop for this short producing time. The contrast between the relatively favorable comparison between derivatives and the poor comparison between the respective pressure changes illustrates that flow-period identification in buildup data must be focused on the shapes of and rules for the derivative.

One further point concerning the buildup pressure rise is worth making. For this short producing time, the buildup pressure rise approaches—and, for late times, is approximately equal to—the drawdown derivative. This is a graphical illustration of the so-called "impulse test" solution to the single-flow and buildup problem (Raghavan 1993). For an ideal impulse test, a small volume of fluid is removed from the well instantaneously, and the subsequent shut-in pressure response contains the data of interest. This is approximated by producing a well for a short time and then focusing the analysis on the long shut-in time buildup response. For example, this testing approach and its analysis are important for pressure falloff data collected after short injection periods. As was discussed in Chapter 1, this is a testing approach used in lower-permeability formations to obtain an estimate of permeability prior to designing and performing fracture stimulation.

2.4.8 Gas Well Testing and Other Complications. Recall that the interpretation and analysis machinery we have built is based firmly on the drawdown paradigm: the initial response from a well producing a fluid with constant properties at a single constant rate from an infinite homogeneous reservoir. In particular, our approach to single-point flow and buildup analysis was to find a way to differentiate and plot the pressure buildup data that allowed us to use the drawdown paradigm machinery.

This idea of manipulating the data collected from more-complex testing situations to make the machinery of the drawdown paradigm applicable is a common theme throughout pressure-transient-analysis theory and application. Outside of the area of interpreting and analyzing data subject to multiple rate changes, multirate drawdown, and buildup testing, the place where this approach has received the most attention and enjoyed the broadest success is in the testing of natural-gas-producing wells.

The main complications that arise when we consider building an approach for gas-well testing are centered on the fact that gas does not have constant fluid properties. Viscosity, density, and compressibility all vary with pressure and temperature. These dependencies on pressure must be incorporated into analysis. To incorporate them, we use two transformations known as the *pseuodopressure* and *pseudotime*.

The real gas pseudopressure, $m(p)$, is usually defined by

$$m(p) = \frac{1}{2} \int_{p_{base}}^{p} \frac{p\,\mathrm{d}p}{\mu z} \quad \dots\dots\dots\dots\dots\dots\dots\dots\dots\dots\dots\dots\dots\dots\dots\dots \quad (2.34)$$

The lower limit of the integral, p_{base}, is an arbitrary value of pressure chosen in a way to make it a convenient reference.

The real gas pseudopressure can be easily shown to partially account for the impact of varying fluid properties. To deal with the remaining nonlinearity, Agarwal (1979) proposed building on the success of the pseudopressure by introducing a second change of variable that he called pseudotime.

Agarwal's pseudotime is defined by

$$t_p = \int_{0}^{t} \frac{\mathrm{d}\tau}{\mu c_g} \quad \dots\dots\dots\dots\dots\dots\dots\dots\dots\dots\dots\dots\dots\dots\dots\dots\dots\dots\dots\dots \quad (2.35)$$

For the single-point-flow and buildup tests that we have focused on, the results of a large body of numerical simulation work on the applicability of these variable transformations to

---

Section Takeaway

- Analyzing data from gas wells relies on transforming data using pseudopressure and pseudotime.
- Based on numerical results, the drawdown paradigm methods can be used on pseudopressure- or pseudotime-transformed data.

gas-well testing can be reduced to two prescriptions for successful use of the drawdown paradigm results. They are:

1. To analyze flow (drawdown) data from a single-point-flow and buildup test, replace pressure with pseudopressure and continue to use time, not pseudotime.
2. To analyze the buildup portion of a single-point-flow and buildup test, replace pressure with pseudopressure and shut-in time with Δt_p.

Most commercial analysis software programs use pseudopressure as the default plotting function for gas-well data. Generally, however, the user is charged with the choice of when and whether to employ the pseudotime transformation.

Before moving on to demonstrate the analysis of a fractured-well pressure buildup test, we will briefly mention two other common complications. The first is the analysis of data complicated by many rate changes during or prior to the period where the data for analysis are collected. The second is accounting for the effects of sealing boundary configuration and location on test data.

Regarding data affected by multirate changes, the more common situation is multirate buildup data (i.e., pressure buildup data collected after a production period in which the rate changes many times as opposed to only once, as in the single-point flow and buildup test, where our interests have been focused). Less common is the call to analyze data collected over a period of production with many rate changes. The exception to this is rate-time decline analysis, generally confined to data from fracture-stimulated tight gas wells, which we will take up after the example buildup analysis. In both instances, the impact of varying rates is accounted for by using a multirate version of Agarwal's equivalent time or another multirate time function from Odeh and Jones (1965). Most commercially available software packages recognize this kind of data and automatically employ one or the other of these time functions as the default plotting approach. The only decision the user need make is how much of the available rate-change data should be entered to garner a reasonable interpretation and analysis. For multirate buildup data, one common rule is to enter prebuildup rate information covering three times the length of the buildup test being analyzed. This rule, however, is based on assumptions about the goals of the analysis; an equally good rule is to just enter all the rate data you have as accurately as you can.

Accounting for the effects of sealing boundary configuration and location on test data requires recognizing the different derivative shapes that result from these effects. A summary of these shapes for ideal single-rate drawdown data can be found in Ehlig-Economides (1988). Some feel for the accuracy of these rules for pressure buildup data plotted as we have recommended can be found in Spivey and Lee (1999).

2.4.9 Example Pressure Buildup Analysis. This is a test performed on a low-permeability gas producer located in western North America. The well was fracture stimulated and then produced for ~150 hours before being shut in for the pressure buildup test. Average production rate was ~1.8 MMscf/D.

An overview plot of the pressure buildup data is shown in **Fig. 2.20.** The measured pressure is shown in red, and the rate is plotted in blue. Note that the length of the shut-in period is more than 300 hours. In addition, a few hours of flowing data are shown before the start of the pressure buildup test period to illustrate the kinds of fluctuations typical of real flowing-pressure data. These variations are among the primary reasons for preferring pressure

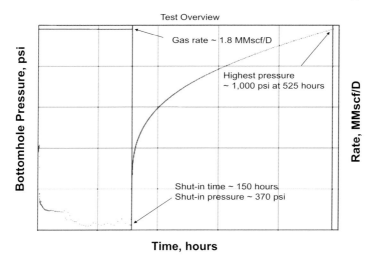

Fig. 2.20—Overview plot: single point flow and buildup example.

buildup tests to drawdown testing. Differentiating such data is clearly not going to give a clean signature such as we had in the simulated drawdown example presented earlier. Despite the fact that the well is clearly not flowing at a single constant rate prior to the pressure buildup period (note the flowing pressure variations indicating this), we shall still analyze the buildup data using a single average rate. This approach is often used in practice and will often give an excellent first approximation to the results of a more detailed analysis incorporating all of the rate changes prior to the test using a more complex time function.

Fig. 2.21 is the log-log diagnostic plot of the pressure buildup data. Since this is a gas-well pressure buildup, the pressure data have first been transformed to pseudopressure using fluid-property data specific for the produced gas. Similarly, time has been converted to pseudotime with the same fluid-property information. Thus, the red points on the plot represent the pseudopressure rise during the buildup test, and the blue points represent the derivative of the pseudopressure rise with respect to Agarwal's equivalent time incorporating pseudotime. Finally, these functions are plotted against elapsed pseudotime. All of this conforms to the rules outlined previously for applying the drawdown paradigm to the interpretation and analysis of gas-well responses and pressure buildup data, respectively. We now can proceed to apply the rules for flow-period identification derived from our understanding of single-rate drawdown testing.

First, the response at early time shows a period in which the pseudopressure and its log-time function derivative overlap. Striking a line of unit slope in these data confirms this to be the wellbore-storage-dominated part of the response. The program we are using in this example automatically does the appropriate wellbore-storage flow-period analysis for us and returns a value for the wellbore-storage coefficient, C, equal to 0.087 bbl/psi.

We know that for a vertically fractured well, we now need to search for three additional flow periods: the bilinear-, linear-, and radial-flow periods. We also know that we need to have an estimate of formation permeability before an estimate of fracture half-length can

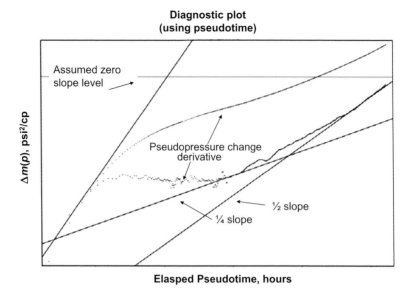

Fig. 2.21—Diagnostic plot: example buildup.

be obtained from the linear flow period, and this half-length estimate is required before we can get a value of fracture conductivity from bilinear-flow data.

Estimating permeability requires a radial flow period, and we expect—according to the drawdown solutions that we have examined—the radial flow period to occur after all the fracture flow periods end. Striking quarter-slope and half-slope lines on the diagnostic plot in Fig. 2.21, we find that the half-slope line fits the derivative data at the end reasonably well, while the only place in the derivative that shows quarter-slope is a short period immediately after the approximately constant derivative period following wellbore storage. Though the bilinear-flow period is almost certainly not correct, we will accept the position of this flow-period identification for now.

With these first approximations to the linear- and bilinear-flow periods chosen, it is evident that the approximately constant derivative period immediately following wellbore storage cannot be the radial period consistent with the present interpretation of the data. It occurs before the fracture flow-periods chosen and, hence, does not conform to the requirements imposed by the drawdown paradigm. We will consider what these data imply about the response shortly.

Eliminating this part of the data leaves us with the obvious conclusion that the measured pressure buildup response was too short to reach the radial-flow period. In light of our earlier comments, the only ways to proceed are to either guess a value of permeability or to use another source of information. Two options for sourcing an estimated value are to look for a permeability value from a core or an estimate from a test on a nearby well. Any estimate derived from a core measurement should be corrected to make it representative of in-situ conditions. In this case, we chose to guess a permeability value by just fitting a constant slope line on the plot somewhere above the end of the data. From the simulated drawdown data we have seen, we know that the derivative will continue to rise up to the correct radial-flow period. We also know that the estimate of permeability depends only on the value (level) of the derivative, not on where it occurs in time. Hence, this approach is clearly reasonable. Since we will adjust all the model-parameter values through history matching, we need not be too precise at this point.

For the constant slope line shown, the program we are using immediately returns an estimated value for permeability. In this case, we get $k = 0.46$ md. With this in hand, we can analyze the data in the linear- and bilinear-flow periods to obtain estimates for x_f and F_{CD}. The interpretation package used here will not give these values directly from the log-log diagnostic plot.

Figs. 2.22 and 2.23 show the linear- and bilinear-flow-period analyses, respectively. These are plots of pseudopressure change vs. square root and fourth root of equivalent time based on pseudotime. For the long producing time of this test, equivalent time based on pseudotime reduces to elapsed pseudotime, making the plot consistent with the approach advocated here. Another time function, the tandem root, is also often used (Swift 1988). The pros and cons of each of these approaches will not be discussed here. In practice, any of these will give reasonable approximate values that can be adjusted to final estimates through history matching. The results from the two plots are $x_f = 264$ ft and $F_{CD} = 3.66$.

Fig. 2.24 shows a comparison of the measured pressure buildup data (blue symbols) with the buildup response computed with the estimated wellbore-storage coefficient, permeability, fracture half-length, and conductivity values. The comparison confirms that the estimates, while not totally incorrect, need to be adjusted.

Though not shown here for reasons of space, a few computations with parameter values across a reasonable range show that the flat slope period early in the data cannot be matched by merely adjusting the values of the four parameters we have estimated. A fifth parameter, fracture skin, is required. This model parameter was discussed briefly in Section 2.4.5. The pattern associated with this parameter is an increase in the spacing between derivative and pressure change. This increase is most pronounced right after the wellbore-storage period,

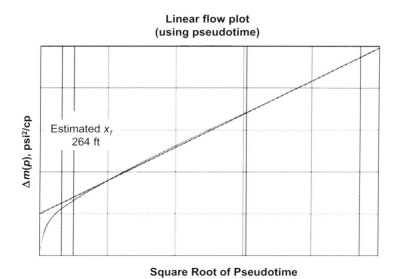

Fig. 2.22—Linear flow plot: example buildup.

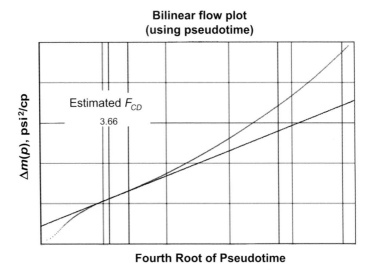

Fig. 2.23—Bilinear flow regime plot: example buildup.

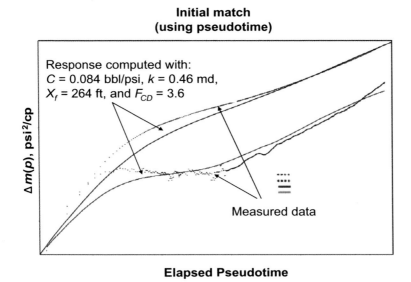

Fig. 2.24—Match using flow-regime-parameter estimates: example buildup.

as the detailed responses showing the impact of fracture skin documented in Cinco-Ley and Samaniego (1977) illustrate.

In any case, an increase in the spacing between the computed pseudopressure rise and its derivative is required if we are to improve the first pass match of the data shown in Fig. 2.24. Introducing a nonzero value for fracture skin, then adjusting all five parameter values with a least-squares-fitting algorithm, yields the match of the measured data shown

Comparison of Final Match and Measured Data

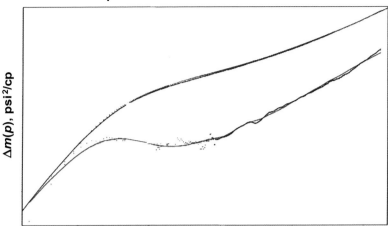

Elasped Pseudotime

Fig. 2.25—Final match with adjusted parameters: example buildup.

in **Fig. 2.25.** Final parameter values obtained are: $C = 0.062$ bbl/psi, $k = 0.16$ md, $s_f = 0.02$, $x_f = 520$ ft, and $F_{CD} = 20$. Note that for this particular pressure buildup response, significant adjustments have been made to all the initial estimates of model parameter values. This reflects both the poor definition of the linear- and bilinear-flow periods and the complete absence of the radial-flow period. While this interpretation is consistent with the pressure buildup data, it certainly has not yielded absolutely certain estimates of the model parameters. How completely the fracture is cleaned up, how accurate the rate data are, and how good our final value of permeability is—all of these impact the accuracy of these parameter estimates. Obtaining parameters and models that allow us to predict long-term performance is certainly the goal of post-appraisal, but, in practice, further adjustments to accepted models and their estimated parameter values must be expected as more performance data become available. We will illustrate this next in Chapter 3.

Chapter 3

Performance Prediction

3.1 Introduction

In this section, we move deeper into the reservoir engineering aspects of hydraulically fractured wells, to rate and recovery prediction. Chapter 2 gave us methods for estimating the in-situ characteristics of the resulting fracture stimulation. Here we will look at methods for using these estimated characteristics to predict the impact of the fracture stimulation on well performance. Some of these methods will also help us further refine the character of the fracture stimulation.

Reservoir engineering is predominantly centered on two activities: calculating how much of a hydrocarbon accumulation can be brought to surface (reserves vs. resources), and calculating the rate at which this can be done. These activities can be done separately, but they are usually done in sequence, with rate estimates providing granularity for and improvement to the reserves estimate. For reservoir engineers, performance prediction means calculating an estimate of production rate as a function of time. This can be done on a well, pattern, field-segment, or fieldwide basis, with each approach having advantages and disadvantages. Here we will look at rate predictions at the single-well scale because this fits best with the themes and inputs of Chapters 1 and 2. Learning to place, post-appraise, and predict performance for a hydraulic-fracturing campaign is very often a horizon-by-horizon, well-by-well process.

In this section, we will look at three approaches for obtaining well-rate predictions: type curves, approximate analytical methods, and full numerical stimulation. For each method, we will first outline the basics of the approach, discussing its assumptions, advantages, and limitations. After this introduction to the three approaches, we will use two of these methods to demonstrate predicting performance for a well-documented case from the literature, where the performance predicted can be readily compared to what the well actually did. The section will close with some general concluding comments.

Prediction is the most important thing that scientists and engineers do. In fact, for many, the ability to make a quantitative prediction of the future state of a system defines what science is. Every other activity should have as its goal providing quality input for performance prediction. Keeping this view in mind, we will start the discussion with type curve methods.

3.1.1 Type Curve Methods. All wells have two common parts to their rate-vs.-time performance. During the first part, the surface rate can be maintained constant while the flowing

---

Section Takeaway

- Plateau and decline periods are common to all well-production profiles.
- Type-curve methods focus on predicting performance during the decline period.
- The Fetkovich type curves are the archetype format. This format breaks the data into transient and boundary-dominated time periods.
- Modifications/additions to the Fetkovich approach include derivatives, time functions, and special type curves for fractured wells.
- Type curve predictions involve three steps: recognizing transient and boundary dominated data sections, fitting history to available type curves, and extrapolating consistently.
- Obtaining rate-time predictions is complicated by time functions.

---

bottomhole pressure declines. This is the time period exploited in Chapter 2 to derive estimates of fracture and reservoir characteristics. This part of a well's rate behavior is called the production plateau period; it can be maintained until the pressure at the sandface is no longer sufficient to push the fluid to surface, at plateau rate, against surface facility operating pressure. The most common ways of extending the plateau period are to inject fluid into the reservoir, maintaining reservoir pressure (and, hence, flowing bottomhole pressure), lowering the pressure in the surface equipment against which the bottomhole pressure has to push the fluid to surface, or stimulating the well, raising the flowing bottomhole pressure by decreasing the pressure drop taken by the flowing fluid as it moves to the well.

No matter what we do, however, eventually the flowing bottomhole pressure will be decreased to the point at which the plateau rate cannot be maintained and the well will enter the second common performance period. Generally, during this period, both the rate and the flowing bottomhole pressure will decline. However, in many instances, artificial-lift systems are installed at this point to keep the flowing bottomhole pressure essentially constant while the rate declines. For example, the majority of oil wells in the United States have pumping systems operated in such a way that they maintain the wells in a pumped-off state (no fluid head above the pump point). This results in a constant bottomhole flowing pressure while the rate declines. Other circumstances with naturally flowing wells often lead to approximately constant flowing bottomhole pressures.

Type curves are aimed at making performance predictions during this second phase of a well's performance; they take as a basic assumption that flowing bottomhole pressure is held constant while well rates decline.

The earliest performance prediction methods in the petroleum industry are based on extrapolating trends in measured data. These rate-decline methods were studied in detail by Arps (1945). Based on empirical observations from a large number of naturally declining producing wells, he arrived at a quantitative expression that covered all of the rate-decline behavior he had observed. This empirically based equation is

$$\frac{q(t)}{q_i} = \frac{1}{(1 + bDt)^{\frac{1}{b}}} \quad \dots\dots\dots\dots\dots\dots\dots\dots\dots\dots\dots\dots\dots\dots\dots\dots\dots\dots \quad (3.1)$$

In Eq. 3.1, q_i is the plateau rate, and $q(t)$ is the rate at any time after decline from the plateau begins. The empirical constants b and D are determined by fitting Eq. 3.1 to observed rates. The constant b is called the decline exponent, while D is the decline percentage. Maintaining the shown separation of the constants on the time term allows this equation to naturally capture the most common form of assumed rate-decline, exponential decline. From the definition of the base of the natural logarithms, e, the limit of Eq. 3.1 as $b \to 0$ becomes (Fetkovich 1980)

$$\frac{q(t)}{q_i} = \exp(-Dt) \quad \cdots\cdots\cdots\cdots\cdots\cdots\cdots\cdots\cdots\cdots\cdots\cdots\cdots\cdots\cdots\cdots\cdots\cdots \quad (3.2)$$

This exponential rate-decline equation also results as the general solution for the boundary-dominated period of the constant-terminal-pressure production analog of the constant-terminal-rate problem (Locke and Sawyer 1975). This plays a central role in developing Fetkovich's type curve method.

The other obvious limit to Eq. 3.1 is obtained by setting $b = 1$. This gives

$$\frac{q(t)}{q_i} = \frac{1}{(1 + Dt)} \quad \cdots\cdots\cdots\cdots\cdots\cdots\cdots\cdots\cdots\cdots\cdots\cdots\cdots\cdots\cdots\cdots\cdots \quad (3.3)$$

The rate-decline behavior governed by Eq. 3.3 is called harmonic decline. This limit of Eq. 3.1 plays a central role for both the Blasingame et al. (1991) and Agarwal et al. (1999) type curves.

The process of establishing a trend on some standard plot and extrapolating it to predict performance is also the essential idea of the rate-time decline type curve methods that have been steadily developed since the early 1980s. Fetkovich (1980) was the first to introduce a type curve approach to predicting performance. He has shown by successful use that his type curve has broad applicability across the spectrum of observed declines and can be used for both damaged (positive skin) and stimulated (negative skin) wells. It is in this sense (i.e., applicability to wells with negative skin) that the Fetkovich type curves can be used for predicting the performance of vertically fractured wells. The work of Fetkovich is seminal to all modern rate-decline type curves. Its most influential descendants are in the publications of Blasingame et al. (1991) and Agarwal et al. (1999).

The idea of a type curve comes from pressure transient theory, and all of the works of Fetkovich, Blasingame et al., and Agarwal et al. are firmly rooted to varying extents in this theory. Pressure transient type curves are log-log plots of typical pressure drop responses predicted for a given well/reservoir model. A number of curves are plotted on the same graph depicting the pressure response across a practical range of model parameters. To make these log-log plots useful, the responses are presented in terms of "dimensionless variables." One typical dimensionless variable, the skin factor, s, was introduced in Chapter 2. Examples of pressure transient type curves can be found in Agarwal et al. (1970) and Bourdet (1983).

The principal difference between pressure transient type curves and rate-time decline type curves lies in what is varying with time and what is held constant as the operating condition. As we have seen, the pressure transient paradigm is that the well pressure varies and the production rate is held constant. For rate-time decline, the paradigm is that the well rate varies and the flowing bottomhole pressure is held constant. This is consistent with the focus of these methods being post-plateau rate prediction.

The Fetkovich type curve is shown in **Fig. 3.1.** As with all rate-time decline type curves, the Fetkovich type curve is a log-log plot of dimensionless rate vs. dimensionless time.

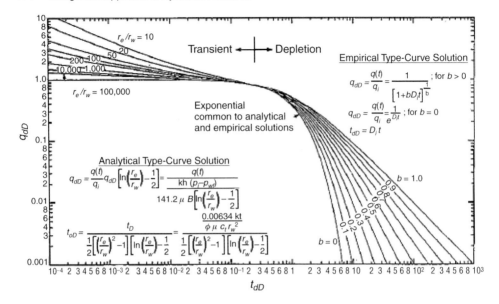

Fig. 3.1—Fetkovich type curve [Fetkovich et al. (1987)].

For an oil well, these quantities are defined by the equations shown on the plot and repeated below for easier reference.

$$q_{dD} = \frac{141.2 B \mu q(t)}{kh(p_i - p_{wf})} \left[\ln\left(\frac{r_e}{r_w}\right) - 0.5 \right] \quad \dots\dots\dots\dots\dots\dots\dots\dots (3.4)$$

$$t_{dD} = \frac{0.00634kt}{\phi(\mu c_t)_i r_w^2} \cdot \frac{1}{0.5\left[\frac{r_e^2}{r_w^2} - 1\right]\left[\ln\left(\frac{r_e}{r_w}\right) - 0.5\right]} \quad \dots\dots\dots\dots\dots\dots (3.5)$$

With the constants shown, the rate is in stock-tank barrels per day, STB/D, and the time is in days. In the dimensionless rate definition, the rate has been explicitly shown as a function of time, while the flowing bottomhole pressure is explicitly shown as a constant. For a well produced against a fixed flowing bottomhole pressure, the producing rate will decline monotonically throughout its life, since the driving force provided by the difference between average reservoir pressure and flowing well pressure will decline monotonically. This is a direct consequence of the finite size of all hydrocarbon reservoirs, and this characteristic is clearly reflected in the shapes of the curves shown in Fig. 3.1.

The main contribution of Fetkovich's work was to establish a quantitative way of amalgamating the theoretically correct transient and boundary-dominated solutions to the constant terminal pressure problem with the broader spectrum of rate-decline behavior embodied in the empirically based generalized rate decline given by Eq. 3.1. This allows a consistent prediction of the well rate across the time period spanned by post-plateau well performance.

To complete the description of Fig. 3.1, we need to look specifically at how some of the individual parameters are defined. The rate-decline curves at early time (transient period) are parameterized by the ratio of effective well drainage radius, r_e, divided by equivalent wellbore radius, r'_w, while at late time (boundary-dominated flow), the controlling parameter is the decline exponent, b. The equivalent wellbore radius was encountered in Chapter 2 and is defined in Eq. 2.27. Effective well drainage radius is defined by

$$r_e = \sqrt{\frac{PV}{\pi \, \overline{h\phi}}}, \qquad\qquad\qquad\qquad\qquad\qquad\qquad\qquad (3.6)$$

where PV is the pore volume connected to the well and the thickness-porosity product in the denominator is the product of average values for these parameters over the connected pore volume.

This type curve includes the impact of stimulation by parameterizing the transient stems by equivalent wellbore radius. Fetkovich shows that the $r'_w = 10$ stem duplicates the rate decline solution for an infinite conductivity vertical fracture produced against a constant terminal pressure. This is consistent with the ability of the wellbore storage and skin solution to explain the pressure transient response for some fractured wells as illustrated in Chapter 2.

The mechanics of using the type curve shown in Fig. 3.1 to predict performance are simple. To give a better understanding of the steps, we assume that this prediction will be made without a software package but with some measured rate-time performance available. First, a copy of Fig. 3.1 is created or obtained and printed out at a convenient scale. The measured well rates vs. time in units consistent with those used in the definitions of the dimensionless variables are then plotted. This plot should be at the same physical scale as the printout of the type curve and is usually printed on a transparency so that the data plot can be overlaid onto the type curve with the type curve visible. The transparent plot of the data is now moved up, down, right and left, maintaining the axes of the data plot parallel to the axes of the type curve, until the best match between a curve on the type curve and the data plot is obtained. A match obtained in this way is shown in **Fig. 3.2.**

If one of the boundary-dominated flow period stems, parameterized by a value of b, is determined by the match, then the future performance of the well can be predicted by transferring the trace of this "depletion" stem directly to the plot of the measured data and reading off rate-time pairs along this trend.

If the data being matched to the type curve do not clearly follow a well-defined "b-stem," as is true for the data in Fig. 3.2, then any rate prediction must be based on rules for choosing a reasonable b-value to determine the decline. Example rules include these: Conservative predictions can always be made by choosing to follow the $b = 0$ stem; a typical gas-well rate decline has $0.3 \le b \le 0.5$, and reservoir layering tends to drive the value of b above 0.5 and toward 1.

While the Fetkovich type curve can be used to make predictions for vertically fractured wells, it works best when fracture lengths are relatively short and fracture conductivities are relatively high. These criteria are more often met by oil-producing wells or water-injection wells.

For fracture gas wells, where fracture lengths tend to be long and conductivities are often low, the Blasingame-McCray (Blasingame et al. 1991) or Agarwal-Gardner (Agarwal et al. 1999) type curves are usually a better choice. These type curves draw their inspiration from

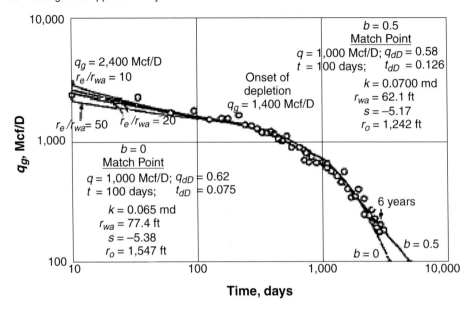

Fig. 3.2—Example type-curve match from Fetkovich et al. (1987).

the Fetkovich approach, but add three improvements. First, both sets of authors present type curves that are specifically geared to finite-conductivity-fracture wells. Second, both sets of authors introduce corresponding derivative curves that help improve the uniqueness of data matching. Finally, both sets of authors include the use of a time transformation appropriate for correlating boundary-dominated data from gas wells with the liquid solutions on which the type curves are based.

These two type curves are very similar in their use, so we choose to discuss only the Agarwal-Gardner approach in some detail. For the Agarwal-Gardner type curve, the dimensionless rate and time functions are now based directly on the pressure transient analytical solutions and are given by

$$q_D = \frac{1422Tq(t)}{kh\left[m(p_i)-m(p_{wf})\right]} \quad \dots\dots\dots\dots\dots\dots\dots\dots\dots\dots\dots\dots\dots\dots \quad (3.7)$$

and

$$t_D = \frac{0.00633kt}{\phi(\mu c_t)_i A} \quad \dots\dots\dots\dots\dots\dots\dots\dots\dots\dots\dots\dots\dots\dots\dots\dots \quad (3.8)$$

In this dimensionless time, the well drainage area, A, plays a role analogous to r_e^2 in Fetkovich's dimensionless time. The Agarwal-Gardner and Blasingame-McCray type curves, like all pressure transient model solutions, are for liquid producing systems. As such, the measured data must be transformed for gas-producing wells to account for fluid-property variations. The use of pseudopressures in Eq. 3.7 fulfills this need during the transient period, but an additional time transformation is required for the data during the boundary-dominated flow period. This time transformation is called *material balance time,* and it is given by

$$t_a = \frac{\mu c_t(p_i)}{q(t)} \int_0^t \frac{q(\tau)\,d\tau}{\mu c_t[\bar{p}(\tau)]}, \quad \dots\dots\dots\dots\dots\dots\dots\dots\dots\dots\dots\dots\dots\dots \quad (3.9)$$

where the notation $\mu c_t[\bar{p}(\tau)]$ means that the viscosity-compressibility product is to be evaluated at average reservoir pressure at the time τ. It is this time transformation that correlates gas-well boundary-dominated data with the liquid solutions during this time frame.

For a gas reservoir at any time, $\bar{p}$ is determined by cumulative production and original gas in place (OGIP). The former will be known, but the latter is generally a quantity to be determined from the type curve match. Hence, determining an Agarwal-Gardner (or Blasingame-McCray) type curve match is an iterative process. An example of an Agarwal-Gardner type curve plot along with its use is shown later in this section.

On the Agarwal-Gardner type curve, the early transient stems are parameterized by F_{CD} and x_e/x_f. This parameterization can be understood intuitively as follows. The underlying solutions used to generate this type curve are those for a fractured well in a square drainage area with characteristic side length x_e. For this case, $A = x_e^2$, implying that x_e is the analog of r_e. Recall also that F_{CD} and x_f determine the equivalent wellbore radius for a vertically fractured well. Thus, we should take the F_{CD} - x_e/x_f combination as the analog of Fetkovich's r_e/r_w'.

Performance prediction with this type curve is similar to the procedure outlined for the Fetkovich curve. If some performance data are available along with an estimate of OGIP, then a plot of rate vs. material balance time can be constructed, and matching this data plot to the best-fit curve determines the position on the type curve and a new estimate of OGIP using the value of A determined from the time match. The match of the early-time data will be easier if estimates of F_{CD} and x_f are available from post-appraisal or design. With this new estimate of OGIP, a new rate-vs.-t_a plot is constructed and rematched to the type curve. These iterations continue until the new and previous values of OGIP agree to an acceptable level. Rate performance prediction can now begin.

Extrapolation along the matched curve into or throughout the only boundary-dominated flow period stem gives the future dimensionless rate performance of the well as a function of material balance time. If we assume that the prediction will be made under a constant-terminal pressure condition, then the dimensionless rate can be readily converted to a real rate value with an estimate of permeability from post-appraisal. Defining the t-t_a relationship is an iterative process.

To do this iteration, we pick a (q, t_a) close to the last measured data point. We now assume the value for the real time, t, and calculate the cumulative production over this time period by the product qt. From material balance, we can now calculate average reservoir pressure. The corresponding material balance time, t_a, is now calculated from the definition or from the following formula given by Agarwal:

$$t_a = \frac{1}{q(t)}(\mu c_t)_i \frac{z_i\,OGIP}{2p_i}[m(p_i) - m(\bar{p})].$$

The calculated value now can be compared to the assumed value. This calculation is repeated until the two values agree. After convergence, the process is repeated for the next chosen rate-material balance time pair. In this way, the full performance prediction can be made. This process is very similar to the semianalytical prediction approach we will discuss in the next part of this section.

If no performance data are available for matching, then post-appraisal or design-based estimates of fracture parameters can be combined with an estimate of OGIP from geologic mapping, offset well data, or analogues to determine the correct curve to predict along. The iterative process then can be followed as described above. Whether done with or without the aid of some data for matching, performance prediction using the Agarwal-Gardner type curves is obviously best done by programming the process on a computer.

The main advantage of the Agarwal-Gardner type curves is that they are tailored for situations in which the transient fracture-dominated period is relatively long. In practice, this most often occurs for low-permeability formations where the placement of long fractures is required to achieve economic performance results. These formations are almost always gas- rather than oil- charged, leading to the single-phase gas slant of much of the literature on these type curve methods.

3.1.2 Analytical/Semianalytical Methods. The main disadvantage of the type curve approaches discussed above is that they are cumbersome for making predictions when future operating conditions vary with time. They are also not geared to making predictions that incorporate the plateau period, when rate is constant and bottomhole pressure is dropping, with predictions for the period when surface operating conditions, fixed or varying, control the natural decline in well rate. Analytical and semianalytical methods allow us to address these two concerns.

This class of prediction methods generally segments into two classes. In the first class, a well inflow performance relationship is combined with the appropriate material balance equation to predict performance. In the second, an analytical solution to the combined transient-boundary dominated flow problem is developed. This solution is invariably a constant-terminal-rate, liquid-flow solution that must be combined using a rate-convolution or numerical-wellbore-coupling approach to allow varying operating conditions. We will look at the first method in some detail. Gao et al. (1994) and Spath et al. (1994) outline the main elements of the second approach. To be concrete, we consider the flow of single-phase gas.

Section Takeaway

- Semianalytical methods come in two forms: inflow-performance relationships combined with material balance and full transient + boundary-dominated solutions combined with rate convolution or numerical wellbore coupling models.
- For fractured wells, the first approach is most suited to fracture stimulations aimed at higher-permeability formations (e.g., TSOs, frac packs).
- Either of these methods can be used to produce plateau + decline period predictions.
- Predictions for complex reservoirs and complex completions can be achieved with these methods.

For the flow of single phase gas, the material balance equation is

$$\frac{G_p(t)}{\text{OGIP}} = \frac{\left(\dfrac{p}{z}\right)_i - \left(\dfrac{p}{z}\right)_t}{\left(\dfrac{p}{z}\right)_i},$$

where $G_p(t)$ is the gas produced at time t. In addition, the p/z ratio in the second term of the numerator on the right side involves the values of average reservoir pressure and z-factor at the same time, t. This equation holds for both the transient and boundary-dominated flow periods.

Inflow performance relationships are essentially equations that relate average reservoir pressure, rate, and flowing bottomhole pressure. Two common inflow relationships for gas are the Rawlins-Schellhardt (1935) equation,

$$q = C\left[\Delta m(p)\right]^n,$$

and the Forchheimer equation (from Dake 1978),

$$\Delta m(p) = Aq + Bq^2.$$

In both of these equations, $\Delta m(p) = m(\bar{p}) - m(p_{wf})$. The constants C and A are related to fundamental model parameters such as k, s, r_e, or A. The constants B and n determine the level of extra pressure drop that occurs when the flow rate is high enough to induce inertial turbulent effects not accounted for by Darcy's law. In practice, all of these constants are most often determined by multirate testing (Dake 1978). However, Carter et al. (1963) give equations for calculating C and n from fundamental model parameters, and Dake (1978) gives expressions for A and B. The main issue that arises when using these equations is estimating the inertial turbulent parameters n or B. In the absence of well-specific data, the best approach is to use information from nearby wells with similar completions.

When either the Rawlins-Schellhardt or Forchheimer equations are used to predict fractured-well performance, the fracture can be incorporated only as an equivalent skin factor calculated from the Cinco-Ley equivalent wellbore radius-F_{CD} relationship shown in Fig. 2.10. This skin is then incorporated into Carter's expression for C or Dake's expression for A. As noted before, this means that the performance during the fracture-dominated flow periods is not accounted for and must be of negligible importance for the prediction. This usually implies that the fractures are relatively short and reasonably conductive, allowing the pseudoradial-flow period to dominate the transient part of the well response. With this stipulation or caution, we generally find that these inflow relationships do a good job of approximating the overall rate performance for most wells. If the fracture-dominated performance period has a significant impact on total well recovery or economic viability, the best approach to performance prediction is the numerical one we will discuss in the next part of this section.

The process for combining material balance with an inflow performance equation is described in the following steps. These steps assume that we have an estimate for OGIP, a table of $m(p)$ vs. p, a value of desired plateau rate, and a value for minimum bottomhole pressure required to lift any given rate against any expected surface operating pressure.

Operating surface pressure can vary with time, but this added complication moves the performance prediction beyond something that can readily be done without programming the process on a computer. The steps described below assume constant bottomhole operating conditions. The steps can be modified for constant surface conditions by adding the calculation of the wellhead pressure from bottomhole pressure to the obvious points in the calculation.

1. Establish the times, t_i, $i = 1, N$, where we would like to predict the rates. These times are often equally spaced at daily or monthly intervals, depending on how rapidly we expect the rate to change. For this type of prediction, this is determined by the magnitude of the expected plateau rate and the size of OGIP. Lower plateau rates can be expected to be sustained longer, and declines from plateau will be more gradual. If OGIP is small, then even low plateau rates will not last, and declines from even low rates will be rapid. Conversely, if OGIP is large, then higher plateau rates can be sustained, and declines from plateau will be dampened when compared to the low OGIP case.

2. For $i = 1$, use the plateau rate, q_{int}, along with the initial pressure to calculate $m(p_{wf})$ from the inflow performance relationship being used. Next, interpolate in the $m(p)$ vs. p table to find p_{wf}.

3. If the calculated bottomhole flowing pressure is above the minimum required to lift the rate to surface, then calculate the cumulative production, $q_{int}t_1$, and then using the material balance equation, calculate the average reservoir pressure at t_1. If the calculated bottomhole flowing pressure is not high enough to lift the plateau rate to surface, then the assumed plateau rate is too high. The plateau rate can be reduced and the steps repeated, or the process for rate calculation after the plateau period, described below, can be used.

4. Continue with Steps 2 and 3 until the predicted bottomhole pressure falls below the corresponding required pressure to lift the plateau rate.

5. When the plateau rate can no longer be sustained, use the average reservoir pressure and the minimum flowing bottomhole pressure in the inflow performance relationship to predict the rate at this point in time. If this time is t_k, calculate the cumulative production $q(t_k)(t_{k+1} - t_k)$.

6. Use the material balance equation to calculate the average reservoir pressure at t_{k+1}.

7. With this new average reservoir pressure, repeat Steps 5 and 6 for the next prediction time.

8. Carry the process on until the prediction is complete for all times.

The approach described above is an analytical performance prediction even though it requires evaluation on the computer for the most general case.

This class of prediction method has been extended to some very complex well/reservoir/fluid systems. Fevang and Whitson (1996), for example, outline a procedure for predicting the depletion performance of gas condensates produced from vertically fractured or horizontal wells.

3.1.3 Numerical Methods. The relative simplicity of the prediction methods presented so far is a great advantage. Input data are limited to a few model parameter values and an estimate of OGIP. Naturally, this is both a strength and a weakness. The full context of detailed reservoir characterization cannot be entered, so the impact of this characterization cannot be studied. In addition, we often want to know what impact changes in the state of the completion will have on future performance. Fracture conductivity and effective fracture length, for example, can improve or degrade with continued production. Numerical methods can more easily and accurately take these kinds of effects into account when matching history or predicting performance.

The standard numerical methods used in the petroleum industry focus on developing complete solutions to the partial differential equations and boundary conditions for the exact problem under consideration. As we saw in Chapter 2, analytical methods more often focus on finding full solutions to approximations of the exact problem. In the limited space we have here, only a short survey of numerical approaches can be accomplished. We will focus on a very general outline of the numerical solution of a simple problem: 2D flow of a liquid in a homogeneous reservoir containing a vertically fractured well. A schematic of this problem is shown in **Fig. 3.3.** In this figure, the fracture is shown as the solid black rectangle at the top of the figure. We assign it a fracture half-length, x_f, a width, w, and proppant permeability and porosity, k_f and x_f. The reservoir has the dimensions, D_x by D_y, with permeability and porosity, k and ϕ. Fluid viscosity and compressibility have the standard symbols. We assume that the well produces at a plateau rate, q, and we must predict

---

Section Takeaway
- Numerical prediction methods are general and flexible.
- Numerical methods have four steps: breaking the reservoir into discrete gridblocks, defining the variables and input parameters for each gridblock, approximating the flow equations in difference form, and solving the resulting system.
- Using general-formulation commercial simulations packages is the industry norm.

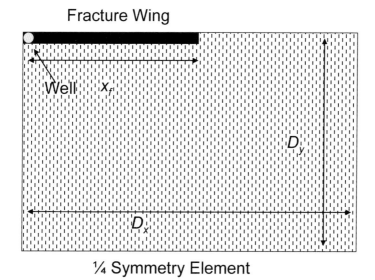

Fig. 3.3—Schematic of example problem.

the bottomhole pressure. After bottomhole pressure falls to a specified level, this pressure is held fixed, and we need to predict the production rate, $q(t)$. At time zero, the pressure is constant throughout the fracture and the reservoir with value p_i. The area shown in Fig. 3.3 is a quarter-symmetry element of the full 2D system. Symmetry is often invoked in solving problems numerically because this allows the use of fewer gridblocks (and, hence, less computation time) to achieve the same accuracy.

In the numerical approach, the region of the reservoir and the region of the fracture are divided into rectangular segments of varying size. These segments are called *gridblocks*. For the problem we are considering here, the gridblocks covering the fracture would be as wide a dimension as the fracture and of varying length (*x*-dimension); there would be small lengths near the well and the fracture tip, with the gridblock lengths increasing as we move away from the well and the fracture tip toward the center of the fracture.

For the reservoir, the gridblocks would start with widths equal to the fracture width and increase as we move away from the well in the y direction. Gridblock lengths for $0 \leq x \leq x_f$ are determined by the lengths assigned in the fracture gridblocks. Lengths for $x_f \leq x \leq D_x$ increase again as we move away from x_f to D_x. Each gridblock is envisioned as having a value of pressure, permeability, saturation, and porosity assigned to it. The rock and fluid properties are assigned directly from the appropriate reservoir or fracture value. The pressures in each gridblock are what we want to calculate.

With gridblocks defined, the partial differential equation governing fluid flow can be *discretized*. This process involves replacing the continuous derivatives in these equations by discrete approximations involving the unknown pressure and the rock and fluid properties in each gridblock. To illustrate, if we look at the unknown pressure in Gridblocks 3, 4, and 5, then the second derivative with respect to x at Gridblock 4 can be approximated by

$$\frac{\partial^2 p}{\partial^2 x}\bigg|_{x_4} \approx \frac{2}{x_4}\left[\frac{(p_5 - p_4)}{(x_5 + x_4)} - \frac{(p_4 - p_3)}{(x_4 + x_3)}\right],$$

where x_3, x_4, and x_5 are the lengths (*x*-dimensions) of Gridblocks 3, 4, and 5 and the subscripted pressures represent the unknown pressures in Gridblocks 3, 4, and 5. This equation represents the standard *central difference formula* for the second derivative.

Substituting these *finite-difference* approximations for the derivatives occurring in the flow equations, along with appropriate values for rock and fluid properties, results in a system of linear equations in the unknown gridblock pressures. These equations now can be solved to obtain the values of the pressure by imposing the relevant production constraint, plateau rate or fixed bottomhole pressure. These values of pressure are computed at a given time in the production history. Values of pressure at a subsequent time are computed by using the pressure at the known time while honoring the production constraint at the subsequent time.

The reader interested in further details of the numerical approach can look at Aziz and Settari (1979), Huyakorn and Pinder (1983), and Peaceman (1978).

3.1.4 Example Application: Performance Prediction.
To partially illustrate some of the methods discussed above, we will outline the prediction of rate performance using the numerical and one of the type curve approaches. This particular example well is taken from Hager and Jones (2001). **Fig. 3.4** shows the final results of the interpretation of a

Log-Log plot (using pseudotime)

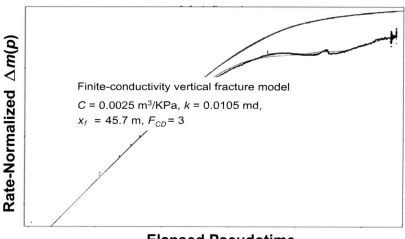

Fig. 3.4—Interpretation of post-fracture-stimulation buildup data from example well.

post-fracture-stimulation pressure-buildup test on a low-permeability gas well. This test was performed within a few days of completing the fracture stimulation.

The main problem with these pressure-buildup results is that the fracture half-length indicated is quite small relative to the size of the fracture expected on the basis of the stimulation design. This is most likely caused by poor fracture cleanup prior to collecting the baseline pressure buildup shown in Fig. 3.4. We would expect that as the well produces, the fracture will clean up, and both effective half-length and (perhaps) effective conductivity will increase.

Fig. 3.5 shows the performance of this well for the first 105 days after the baseline pressure-buildup test was done. The well rate is shown as black triangles, while the open

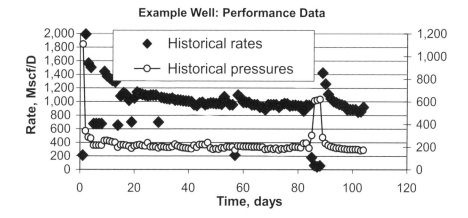

Fig. 3.5—Example well post-fracture performance, first 105 days.

symbols are the flowing wellhead pressure. The axis on the right of the plot gives the value of this quantity. The approach we will follow is to use a numerical simulator, with input determined from the post-fracture buildup test, to predict this 105 days of performance. Since the pressure buildup does not see any boundaries, drainage area will be assumed at 160 acres. Admittedly, this is an arbitrary choice, but we will test this during the prediction exercise.

Our goal is two-fold. First, we would like to determine whether the performance data indicate any improvement in fracture half-length or conductivity during this period. Second, we would then like to predict the future performance of this well on the basis of these adjusted parameter values.

Fig. 3.6 shows a match of the first 105 days of performance data using a commercial gas-reservoir-simulation package. In this simulation, the measured flowing wellhead pressures have been imposed as the operating control, and the well rates have been predicted. The match shown required that the effective fracture half-length be increased from the initial value of 46 m to a final value of 215 m. An increase in fracture half-length was imposed whenever the simulation predicted rate began to fall below the measured rate. To match the entire 105 days of performance data, the fracture half-length had to be increased three times, with the last increase occurring at approximately 60 days. The value of F_{CD} was essentially honored by adjusting the proppant permeability to keep the dimensionless conductivity constant as fracture half-length was increased. Reservoir effective permeability also had to be increased by approximately 10% to achieve the match.

To make a rate prediction, we now need only run the simulator forward in time with some choice of operating condition (e.g., producing against the historical value of wellhead pressure). However, for this well, the above analysis was originally done after the well had already produced for almost 1,000 days. Thus, we will look at the prediction of well rates, imposing the known operating wellhead pressures, holding model parameters constant at the adjusted values obtained from the match of the first 105 days of rate performance.

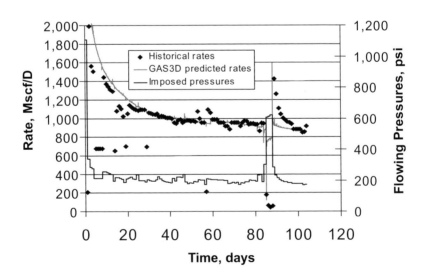

Fig. 3.6—Simulator match: example well performance k = 0.0115 md, effective half-length increasing with time.

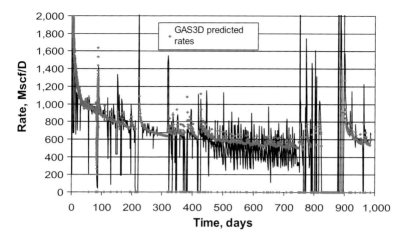

Fig. 3.7—Prediction of subsequent performance based on adjusted parameter values.

Fig. 3.7 shows a comparison of the simulation predicted rates (gray symbols) with the measured well rates for the remainder of the known performance. Obviously the prediction is excellent, indicating that no further adjustment to the parameters is required to explain the well rates. The rates measured between approximately 400 and 750 days are affected by liquid loading, but the average "on-time" performance is still captured quite well by the simulator. While this type of prediction is somewhat unusual, it does illustrate the power of integrating typical post-appraisal data analysis with performance simulation. We will now take a more conventional approach to prediction using a type curve match.

The Agarwal-Gardner approach to type curve prediction is discussed in detail in Agarwal et al. (1999). As shown there, the first step in this approach is to obtain an estimate of OGIP. This is done on a specialized type curve based on a Cartesian plot of dimensionless rate vs. dimensionless cumulative production. On this type curve, the value of OGIP is adjusted until the measured data fall on a straight line that extrapolates through $\pi/2$. The assumption behind this plot is that the data falling on this line are in the boundary-dominated flow period and, hence, are valid data for determining OGIP. No form of diagnostic plotting is inherent in the Agarwal-Gardner approach that confirms this assumption other than the fact that for some guess of OGIP, the late-time data fall on the expected straight line. **Fig. 3.8** shows this analysis for the first 420 days of production data for our example well. This plot was taken from Hager and Jones (2001), but the analysis done here mimics the original analysis documented internally by the source company.

With the reservoir size (i.e., OGIP) determined, the data can be matched to a rate-time type curve, and a rate performance prediction can be obtained.

The best match of the data to one of the forms of the Agarwal-Gardner rate-time type curve, appropriate for fractured wells is shown in **Fig. 3.9.** The data have been matched in a way that preserves, as closely as possible, the estimated permeability and fracture half-length obtained from the transient analysis. Note, however, that unlike the numerical match obtained earlier, the permeability has had to be increased by more than a factor of 2 (from 0.011 to 0.026 md). This is partially a consequence of the limited number of curves there are to match, but it does reflect a typical discrepancy seen between type-curve-derived values and those sourced from PTA and numerical history matching.

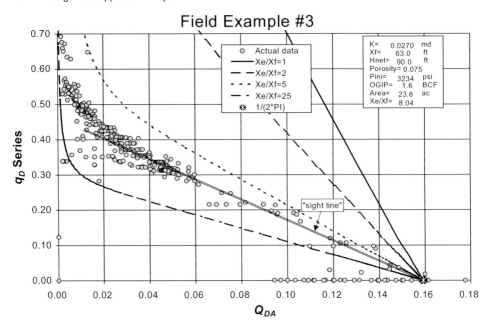

Fig. 3.8—Agarwal-Gardner Cartesian type curve for estimating OGIP [from Agarwal et al. (1999)].

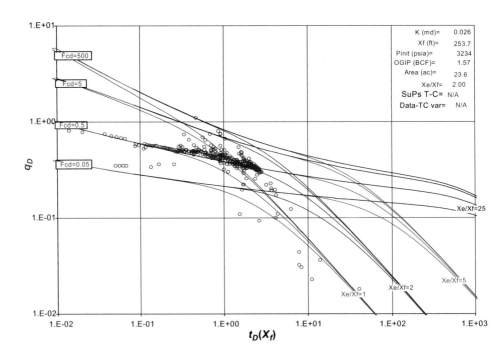

Fig. 3.9—Agarwal-Gardner type-curve match of example data.

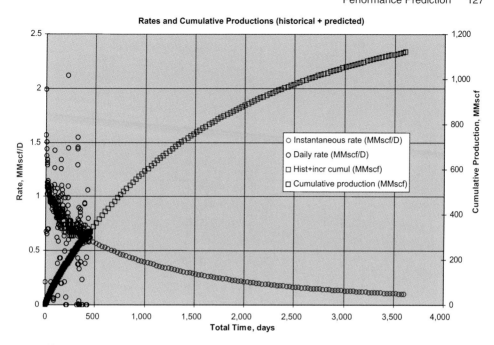

Fig. 3.10—Agarwal-Gardner type-curve rate and cumulative production prediction.

A second issue with this type-curve match is that it is inconsistent with the assumption of boundary-dominated flow used to obtain the OGIP estimate in Fig. 3.8. Very little of the data we would take as reliable (i.e., ignoring the scatter of outliers below and to the right of the main band of data) is actually deep into boundary dominated. There are no data on the negative unit slope trend of the type curve. Hence, contrary to what a cursory examination of Fig. 3.8 would imply, this match shows that the well is still essentially in transient flow. This immediately implies that any rate prediction we make with the assumed OGIP will be conservative. The drainage area implied by this OGIP value is approximately 23 acres, so the Agarwal-Gardner rate prediction will be considerably more conservative than any we would obtain from the numerical approach used above, since the numerical approach used a 160-acre drainage area. In addition, there is no reason to adjust this area in the numerical prediction, since the numerical prediction reliably predicted the first 1,000 days of production without the need to change the assumed area.

The above discussion reflects our general experience with type-curve-based predictions. They are often conservative relative to numerically based predictions. As is the case here, this can be expected to happen even when more data are used to determine the best type-curve match.

To complete this example, the rate and cumulative production predicted using the Agarwal -Gardner type-curve approach are shown in **Fig. 3.10.** The mechanical process used to obtain this prediction assumes that the flowing pressure controlling the future rates is the final measured flowing pressure in the historical data. Some feel for how conservative this rate prediction is can be obtained by comparing the predicted rate here and the measured rate at ~1,000 days. While the well actually produced approximately 600 Mscf/D at this time, the predicted rate from Agarwal-Gardner is well below 500 Mscf/D. If the correct drainage area is at least 8 times larger than that estimated from the type-curve analysis, as implied by the numerical match, then the difference between actual performance and predicted performance will certainly diverge more with increasing time.

Nomenclature

A = fracture area or drainage area, ft^2 or Darcy flow coefficient in Forchheimer equation, psi^2/cp/Mscf/D

B = formation volume factor, reservoir volume/surface volume; Arps decline exponent; turbulent flow coefficient in Forchheimer equation, psi^2/cp/(Mscf/D)^2

c_{eff} = effective system compressibility, psi^{-1}

c_g = gas-phase compressibility, psi^{-1}

c_o = oil-phase compressibility, psi^{-1}

c_r = formation compressibility, psi^{-1}

c_t = total system compressibility, psi^{-1}

c_w = water-phase compressibility, psi^{-1}

C = fluid-loss coefficient, ft/min$^{1/2}$; pressure decline parameter, ft; wellbore storage coefficient, bbl/psi; Rawlins-Schellhardt coefficient, Mscf/D/psi^n

C_A = Dietz shape factor

C_p = perforation discharge coefficient, ft/min$^{1/2}$

C_t = total fluid-loss coefficient, ft/min$^{1/2}$

d = perforation diameter, in.

D = Arps decline percentage

D_x = reservoir extent in x-direction, ft

D_y = reservoir extent in y-direction, ft

e_f = fluid efficiency

E = Young's modulus, psi

E' = plane strain modulus, psi

Ei = exponential integral

F_{CD} = dimensionless fracture conductivity

G = G-function or pressure-difference function

$G_p(t)$ = gas produced at time t, Mscf

h, H = fracture height or formation thickness, ft

h_p = permeability fracture height, ft

k = formation permeability, md

k_f = proppant permeability, md

K = constant in Eq. 1.18

$K_{k\text{-app}}$ = apparent fracture toughness, psi/in^2

L or L_f = tip-to-tip fracture length, ft

m = semilog slope, psi/log cycle

$m(p)$ = real gas pseudopressure, psi^2/cp

n = non-Newtonian power-law exponent or Rawlins-Schellhardt turbulence coefficient

N = number of perforations or total number of time nodes

p = pressure, psi

p_{base} = base pressure used in definition of $m(p)$, psi

p_{fbhp} = final flowing bottomhole pressure, psi

p_{fnb} = final net treating pressure, psi

p_i = initial pressure, psi

p_k = kth gridblock pressure, psi

p_{net} = net pressure, psi

p_s = net treating pressure at shut-in, psi

p_{tec} = tectonic stress, psi

p_w = well pressure, psi

p_{wb} = pressure measured in the wellbore, psi

p_{wf} = flowing bottomhole pressure, psi

$p_{wf,s}$ = flowing wellbore pressure at the instant of shut-in, psi

p_{ws} = shut-in wellbore pressure, psi

q = pump rate or production rate, B/D (liquid), Mscf/D (gas)

q_D = Agarwal or Blasingame dimensionless rate

q_{dD} = dimensionless decline rate

q_i = initial rate, B/D (liquid) or Mscf/D (gas)

q_{int} = plateau rate, Mscf/D

r or R = radius of radial fracture, ft or radial distance, ft

r_e = well drainage radius, ft

R_{max} = distance to proppant bridging point, ft

r_w = wellbore radius, ft

r_w' = effective wellbore radius, ft

s = fracture stiffness, psi/ft or skin factor

s_f = fracture face skin

S_p = fluid loss per unit area before filter-cake formation, ft

t = elapsed time, minutes or hours

t_a = material-balance time, hours

t_c = fracture-closure time, minutes

t_D = Agarwal or Blasingame dimensionless time

t_{dD} = dimensionless time

t_{eq} = Agarwal's equivalent time, hours

t_i = ith time node

t_p	=	pumping time, minutes; producing time, hours; Agarwal's real gas pseudotime, hr-psi/cp
V_f	=	fracture volume, ft³
V_L	=	lost or leakoff volume, ft³
V_p	=	pumped volume, ft³
w	=	fracture width, ft
x_f	=	fracture half-length, ft
x_k	=	kth gridblock length in the x-direction, ft
z	=	real gas deviation factor
Δp^*	=	pressure drop after pump shutdown, psi
Δp_{DD}	=	drawdown pressure drop solution, psi
Δp_{pf}	=	pressure drop caused by perforation friction, psi
Δp_{skin}	=	pressure drop caused by skin, psi
Δp_{SPFB}	=	single-point flow and buildup pressure drop solution, psi
Δp_{wb}	=	pressure drop in the wellbore, psi
Δt	=	time measured from beginning of pressure buildup, hours
γ	=	Euler's constant
η	=	diffusivity, md-psi/cp
μ	=	fluid viscosity, cp
$(\mu c_t)_I$	=	viscosity-total compressibility product evaluated at initial pressure, cp/psi
ν	=	Poisson's ratio
ρ	=	fluid density, lbm/ft³ or G-function analysis parameter (Eq. 1.16)
σ_c	=	confining stress, psi
$\sigma_{h,\max}$	=	maximum horizontal stress
$\sigma_{h,\min}$	=	minimum horizontal stress
σ_{ob}	=	overburden stress, psi
ϕ	=	porosity

References

Abass, H., Brumley, J., and Venditto, J. 1994. Oriented Perforations—A Rock Mechanics View. Paper SPE 28555 presented at the SPE Annual Technical Conference and Exhibition, New Orleans, 25–28 September. DOI: 10.2118/28555-MS.

Abbaszadeh, M. and Kamal, M.M. 1988. Automatic Type Curve Matching for Well Test Analysis. *SPEFE* **3** (3): 567–577. SPE-16443-PA. DOI: 10.2118/16443-PA.

Agarwal, R.A. 1979. "Real Gas Pseudo-Time"—A New Function for Pressure Buildup Analysis of MHF Gas Wells. Paper SPE 8279 presented at the SPE Annual Technical Conference and Exhibition, Las Vegas, Nevada, 23–26 September. DOI: 10.2118/8279-MS.

Agarwal, R.G. 1980. A New Method To Account for Producing Time Effects When Drawdown Type Curves Are Used To Analyze Pressure Buildup and Other Test Data. Paper SPE 9289 presented at the SPE Annual Technical Conference and Exhibition, Dallas, 21–24 September. DOI: 10.2118/9289-MS.

Agarwal, R.G., Al-Hussainy, R., and Ramey, H.J. Jr. 1970. An Investigation of Wellbore Storage and Skin Effect in Unsteady Liquid Flow: I. Analytical Treatment. *SPEJ* **10** (3): 279–290; *Trans.,* AIME, **249.** SPE-2466-PA. DOI: 10.2118/2466-PA.

Agarwal, R.G., Gardner, D.C., Kleinsteiber, S.W., and Fussell, D.D. 1999. Analyzing Well Production Data Using Combined Type Curve and Decline-Curve Analysis Concepts. *SPEREE* **2** (5): 478–486. SPE-57916-PA. DOI: 10.2118/57916-PA.

Anderson, A.J., Ashton, P.J.N., Lang, J., and Samuelson, M.L. 1996. Production Enhancement Through Aggressive Flowback Procedures in the Codell Formation. Paper SPE 36468 presented at the SPE Annual Technical Conference and Exhibition, Denver, 6–9 October. DOI: 10.2118/36468-MS.

Anderson, T.O. and Stahl, E.J. 1967. A Study of Induced Fracturing Using an Instrumental Approach. *JPT* **19** (2): 261–67; *Trans.,* AIME, **240.** SPE-1537-PA. DOI: 10.2118/1537-PA.

Arps, J.J. 1945. Analysis of Decline Curves. *Trans.,* AIME, **160:** 228–247.

Ayoub, J.A., Hutchins, R.D., van de Bas, F., Cobianco, S., Emiliani, C.N., Glover, M. et al. 2006. New Findings in Fracture Cleanup Change Common Industry Perceptions. Paper SPE 98746 presented at the SPE International Symposium and Exhibition on Formation Damage Control, Lafayette, Louisiana, 15–17 February. DOI: 10.2118/98746-MS.

Aziz, K. and Settari, A. 1979. *Petroleum Reservoir Simulation.* Essex, UK: Applied Science Publishers.

Babcock, E.A. 1978. Measurement of subsurface fractures from dipmeter logs. *AAPG Bull.* **62** (7): 1111–1126.

Bale, A., Smith, M.B., and Settari, A. 1994. Post-Frac Productivity Calculation for Complex Reservoir/Fracture Geometry. Paper SPE 28919 presented at the European Petroleum Conference, London, 25–27 October. DOI: 10.2118/28919-MS.

Barree, R.D. and Mukherjee, H. 1996. Determination of Pressure Dependent Leakoff and Its Effect on Fracture Geometry. Paper SPE 36424 presented at the SPE Annual Technical Conference and Exhibition, Denver, 6–9 October. DOI: 10.2118/36424-MS.

Bell, J.S. and Gough, D.I. 1979. Northeast-southwest compressive stress in Alberta: Evidence from oil wells. *Earth and Planetary Science Letters* **45** (2): 475–482. DOI:10.1016/0012-821X(79)90146-8.

Biot, M.A. and Medlin, W.L. 1985. Theory of Sand Transport in Thin Fluids. Paper SPE 14468 presented at the SPE Annual Technical Conference and Exhibition, Las Vegas, Nevada, 22–26 September. DOI: 10.2118/14468-MS.

Blanton, T.L. 1983. The Relation Between Recovery Deformation and In-Situ Stress Magnitudes. Paper SPE 11624 presented at the SPE/DOE Low-Permeability Gas Reservoirs Symposium, Denver, 14–16 March. DOI: 10.2118/11624-MS.

Blanton, T.L. and Teufel, L.W. 1983. A Field Test of the Strain Recovery Method of Stress Determination in Devonian Shales. Paper SPE 12304 presented at the SPE Eastern Regional Meeting, Champion, Pennsylvania, 9–11 November. DOI: 10.2118/12304-MS.

Blasingame, T.A, McCray, T.L., and Lee, W.J. 1991. Decline Curve Analysis for Variable Pressure Drop/Variable Flowrate Systems. Paper SPE 21513 presented at the SPE Gas Technology Symposium, Houston, 22–24 January. DOI: 10.2118/21513-MS.

Bourdet, D.L., Whittle, T.M., Douglas, A.A., and Pirard, Y.M. 1983. A new set of type curves simplifies well test analysis. *World Oil* (May 1983): 95–106.

Britt, L.K., Hager, C.J., and Thompson, J.W. 1994. Hydraulic Fracturing in a Naturally Fractured Reservoir. Paper SPE 28717 presented at the SPE International Petroleum Conference and Exhibition of Mexico, Veracruz, Mexico, 10–13 October. DOI: 10.2118/28717-MS.

Britt, L.K., Smith, M.B., Haddad, Z., and Lawrence, P. 2006. A Multidisciplinary Approach to Hydraulic Fracturing in the South Texas Wilcox Formation. Paper SPE 102226 presented at the SPE Annual Technical Conference and Exhibition, San Antonio, Texas, 24–27 September. DOI: 10.2118/102226-MS.

Britt, L.K. 1985. Optimized Oilwell Fracturing of Moderate-Permeability Reservoirs. Paper SPE 14371 presented at the SPE Annual Technical Conference and Exhibition, Las Vegas, Nevada, 22–26 September. DOI: 10.2118/14371-MS.

Brown, R.O., Forgotson, J.M., and Forgotson, J.M. Jr. 1980. Predicting the Orientation of Hydraulically Created Fractures in the Cotton Valley Formation of East Texas. Paper SPE 9269 presented at the SPE Annual Technical Conference and Exhibition, Dallas, 21–24 September. DOI: 10.2118/9269-MS.

Carter, R.D., Miller, S.C. Jr., and Riley, H.G. 1963. Determination of Stabilized Gas Well Performance From Short Flow Tests. *JPT* **15** (6): 651–658; *Trans.*, AIME, **228.** SPE-400-PA. DOI: 10.2118/400-PA.

Castillo, J.L. 1987. Modified Fracture Pressure Decline Analysis Including Pressure-Dependent Leakoff. Paper SPE 16417 presented at the Low-Permeability Reservoirs Symposium, Denver, 18–19 May. DOI: 10.2118/16417-MS.

Cinco-Ley, H. and Samaniego, V. 1977. Effect of Wellbore Storage and Damage on the Transient Pressure Behavior of Vertically Fractured Wells. Paper SPE 6752 presented at the SPE Annual Technical Conference and Exhibition, Denver, 9–12 October. DOI: 10.2118/6752-MS.

Cinco-Ley, H. and Samaniego-V., F. 1981. Transient Pressure Analysis for Fractured Wells. *JPT* **33** (9): 1749–1766. SPE-7490-PA. DOI: 10.2118/7490-PA.

Cinco-Ley, H., Samaniego, V.F., and Dominguez, A.N. 1978. Transient Pressure Behavior for a Well With a Finite-Conductivity Vertical Fracture. *SPEJ* **18** (4): 253–264. 10.2118/6014-PA.

Clark, J.B. 1949. A Hydraulic Process for Increasing the Productivity of Oil Wells. *Trans., AIME,* **186:** 1–8.

Clark, R.C. et al. 1953. Application of hydraulic fracturing to the stimulation of oil and gas production. *Drill. & Prod. Prac.,* API, 113–22.

Cooke, C.E. Jr. 1973. Conductivity of Fracture Proppants in Multiple Layers. *JPT* **25** (9): 1101–1107; *Trans.,* AIME, **255.** SPE-4117-PA. DOI: 10.2118/4117-PA.

Cooper, G.D., Nelson, S.G., and Schopper, M.D. 1984. Comparison of Methods for Determining In-Situ Leakoff Rate Based on Analysis With an On-Site Computer. Paper SPE 13223 presented at the SPE Annual Technical Conference and Exhibition, Houston, 16–19 September. DOI: 10.2118/13223-MS.

Dake, L.P. 1978. *Fundamentals of Reservoir Engineering.* Amsterdam: Elsevier Science.

Davis, P.M. 1983. Surface deformation associated with dipping hydrofracture. *J. Geophysical Research* **881** (B7): 5826–5234. DOI:10.1029/JB088iB07p05826.

Dobkins, T.A. 1981a. Improved Methods To Determine Hydraulic Fracture Height. *JPT* **33** (4): 719–726. SPE-8403-PA. DOI: 10.2118/8403-PA.

Dobkins, T.A. 1981b. Procedures, Results, and Benefits of Detailed Fracture Treatment Analysis. Paper SPE 10130 presented at the SPE Annual Technical Conference and Exhibition, San Antonio, Texas, 4–7 October. DOI: 10.2118/10130-MS.

Dutton, R.E., Nolte, K.G., and Smith, M.G. 1982. Use of the Long-Spaced-Digital-Sonic Log to Determine Relationships of Fracturing Pressure and Fracture Height for Wells in the East Texas Cotton Valley Tight Gas Play. Company Report F82-P-12, Amoco Production Company, Houston (15 February 1982).

Ehlig-Economides, C. 1988. Use of the Pressure Derivative for Diagnosing Pressure-Transient Behavior. *JPT* **40** (10): 1280–1282. SPE-18594-PA. DOI: 10.2118/18594-PA.

Elbet, J.L., Howard, R.L., Talley, G.R., and McLaughlin, B.H. 1984. Stimulation Study of Cottage Grove Formation. *JPT* **36** (7): 1199–1205. SPE-11564-PA. DOI: 10.2118/11564-PA.

Ely, J.W. 1985. *Stimulation Treatment Handbook: An Engineer's Guide to Quality Control.* Tulsa: PennWell Publishing Company.

Farris, R.F. 1953. Fracturing formations in wells. US Patent No. RE23,733 (10 November 1953).

Fetkovich, M.J. 1980. Decline Curve Analysis Using Type Curves. *JPT* **32** (6): 1065–1077. SPE-4629-PA. DOI: 10.2118/4629-PA.

Fetkovich, M.J., Vienot, M.E., Bradley, M.D., Kiesow, U.G. 1987. Decline Curve Analysis Using Type Curves: Case Histories. *SPEFE* **2** (4): 637–656. SPE-13169-PA. DOI: 10.2118/13169-PA.

Fevang, Ø. and Whitson, C.H. 1996. Modeling Gas-Condensate Well Deliverability. *SPERE* **11** (4): 221–230. SPE-30714-PA. DOI: 10.2118/30714-PA.

Gao, C., Jones, J.R., Raghavan, R., and Lee, W.J. 1994. Responses of Commingled Systems With Mixed Inner and Outer Boundary Conditions Using Derivatives. *SPEFE* **9** (4): 264–271. SPE-22681-PA. DOI: 10.2118/22681-PA.

Geertsma, J. and de Klerk, F. 1969. A Rapid Method of Predicting Width and Extent of Hydraulic Induced Fractures. *JPT* **21** (12): 1571–1581; *Trans.,* AIME, **246.** SPE-2458-PA. DOI: 10.2118/2458-PA.

Gidley, J.L., Holditch, S.A., Nierode, D.E., and Veatch, R.W. Jr. 1989. *Recent Advances in Hydraulic Fracturing*. Monograph Series, SPE, Richardson, Texas **12.**

Godbey, J.K. and Hodges, H.D. 1958. Pressure Measurements During Fracturing Operations. *Trans.*, AIME, **213:** 65–69.

Gough, D.I. and Bell, J.S. 1981. Stress orientations from oil well fractures in Alberta and Texas. *Canadian Journal of Earth Sciences* **18:** 638–645.

Gruesbeck, C. and Collins, R.E. 1982. Particle Transport Through Perforations. *SPEJ* **22** (6): 857–865. SPE-7006-PA. DOI: 10.2118/7006-PA.

Guppy, K.H., Cinco-Ley, H., and Ramey, H.J. Jr. 1982a. Pressure Buildup Analysis of Fractured Wells Producing at High Flow Rates. *JPT* **34** (11): 2656–2666. SPE-10178-PA. DOI: 10.2118/10178-PA.

Guppy, K.H., Cinco-Ley, H., Ramey H.J. Jr., and Samaniego-V., F. 1982b. Non-Darcy Flow in Wells With Finite-Conductivity Vertical Fractures. *SPEJ* **22** (5): 681–698. SPE-8281-PA. DOI: 10.2118/8281-PA.

Hager, C.J. and Jones, J.R. 2001. Analyzing Flowing Production Data With Standard Pressure Transient Methods. Paper SPE 71033 presented at the SPE Rocky Mountain Petroleum Technology Conference, Keystone, Colorado, 21–23 May. DOI: 10.2118/71033-MS.

Holditch, S.A. 1979. Factors Affecting Water Blocking and Gas Flow From Hydraulically Fractured Gas Wells. *JPT* **31** (12): 1515–1524. SPE-7561-PA. DOI: 10.2118/7561-PA.

Holditch, S.A. and Morse, R.A. 1976. The Effects of Non-Darcy Flow on the Behavior of Hydraulically Fractured Gas Wells. *JPT* **28** (10): 1169–1179. SPE-5586-PA. DOI: 10.2118/5586-PA.

Horner, D.R. 1951. Pressure build-up in wells. *Proc.*, Third World Petroleum Congress, The Hague, Sec. II, 503–523.

Hubbert, M.K. and Willis, D.G. 1957. Mechanics of Hydraulic Fracturing. *Trans.*, AIME, **210:** 153–168.

Huitt, J.L. 1960. Hydraulic Fracturing With Single Point Entry Technique. *JPT* **12** (3): 11–13. SPE-1469-G. DOI: 10.2118/1469-G.

Huitt, J.L. and McGlothin, B.B. Jr. 1958. The propping of fractures in formations susceptible to propping-sand embedment. *Drill. & Prod. Prac.* **115.**

Huitt, J.L., McGlothin, B.B. Jr., and McDonald, J.F. 1958. The propping of fractures in formations in which propping sand crushes. *Drill. & Prod. Prac.* **115.**

Huyakorn, P.S. and Pinder, G.F. 1983. *Computational Methods in Subsurface Flow*. New York: Academic Press.

Jin, L. and Penny, G.S. 2000. A Study of Two-Phase, Non-Darcy Gas Flow Through Proppant Pacs. *SPEPF* **15** (4): 247–254. SPE-66544-PA. DOI: 10.2118/66544-PA.

Khristianovich, S.A. and Zheltov, Y.P. 1955. Formation of vertical fractures by means of highly viscous liquid. *Proc.*, Fourth World Petroleum Congress, Rome, Sec. II, 579–586.

Krumbein, W.C. and Sloss, L.L. 1963. *Stratigraphy and Sedimentation*, second edition. New York: Books in Geology, W.H. Freeman and Company.

Lacy, L.L. 1987. Comparison of Hydraulic-Fracture Orientation Techniques. *SPEFE* **2** (1): 66–76; *Trans.*, AIME, **283.** SPE-13225-PA. DOI: 10.2118/13225-PA.

Locke, C.D. and Sawyer, W.K. 1975. Constant Pressure Injection Test in a Fractured Reservoir—History Match Using Numerical Simulation and Type Curve Analysis. Paper SPE 5594 presented at the SPE Annual Meeting, Dallas, 28 September–1 October. DOI: 10.2118/5594-MS.

Maly, J.W. and Morton, T.E. 1951. Selection and evaluation of wells for hydrafrac treatment. *Oil & Gas Journal* **52** (3 May 1951): 126.

Martins, J.P. and Harper, T.R. 1985. Minifrac Pressure Decline Analysis for Fractures Evolving From Long Perforated Intervals and Unaffected by Confining Strata. Paper SPE 13869 presented at the SPE/DOE Low-Permeability Gas Reservoirs Symposium, Denver, 19–22 March. DOI: 10.2118/13869-MS.

Martins, J.P., Abel, J.C., Dyke, C.G., Michel, C.M., and Stewart, G. 1992. Deviated Well Fracturing and Proppant Production Control in the Prudhoe Bay Field. Paper SPE 24858 presented at the SPE Annual Technical Conference and Exhibition, Washington, DC, 4–7 October. DOI: 10.2118/24858-MS.

Martins, J.P., Leung, K.H., Jackson, M.R., Stewart, D.R., and Carr, A.H. 1992. Tip Screen-out Fracturing Applied to the Ravenspurn South Gas Field Development. *SPEPE* **7** (3): 252–258. SPE-19766-PA. DOI: 10.2118/19766-PA.

May, E.A., Britt, L.K., and Nolte, K.G. 1997. The Effect of Yield Stress on Fracture Fluid Cleanup. Paper SPE 38619 presented at the SPE Annual Technical Conference and Exhibition, San Antonio, Texas, 5–8 October. DOI: 10.2118/38619-MS.

May, E.A., Britt, L.K., and Nolte, K.G. 1997. The Effect of Yield Stress on Fracture Fluid Cleanup. Paper SPE 38619 presented at the SPE Annual Technical Conference and Exhibition, San Antonio, Texas, 5–8 October. DOI: 10.2118/38619-MS.

McLennan, J.D. and Roegiers, J.C. 1982. How Instantaneous Are Instantaneous Shut-In Pressures? Paper SPE 11064 presented at the SPE Annual Technical Conference and Exhibition, New Orleans, 26–29 September. DOI: 10.2118/11064-MS.

Medlin, W.L., Sexton, J.H., and Zumwalt, G.L. 1985. Sand Transport Experiments in Thin Fluids. Paper SPE 14469 presented at the SPE Annual Technical Conference and Exhibition, Las Vegas, Nevada, 22–26 September. DOI: 10.2118/14469-MS.

Miller, W.K. II and Smith, M.B. 1989. Reanalysis of the MWX-Fracture Stimulation Data From the Paludal Zone of the Mesaverde Formation. Paper SPE 19772 presented at the SPE Annual Technical Conference and Exhibition, San Antonio, Texas, October. DOI: 10.2118/19772-MS.

Milton-Tayler, D. 1993. Realistic Fracture Conductivities of Propped Hydraulic Fractures. Paper SPE 26602 presented at the SPE Annual Technical Conference and Exhibition, Houston, 3–6 October. DOI: 10.2118/26602-MS.

Morris, C.W. and Sinclair, A.R. 1984. Evaluation of Bottomhole Treatment Pressure for Geothermal Well Hydraulic Fracture Stimulation. *JPT* **36** (5): 829–836. SPE-11157-PA. DOI: 10.2118/11157-PA.

Nolte, K.G. 1979. Determination of Fracture Parameters from Fracturing Pressure Decline. Paper SPE 8341 presented at the SPE Annual Technical Conference and Exhibition, Las Vegas, Nevada, 23–26 September. DOI: 10.2118/8341-MS.

Nolte, K.G. 1986a. Determination of Proppant and Fluid Schedules From Fracturing-Pressure Decline. *SPEPE* **1** (4): 255–265; *Trans.,* AIME, **281.** SPE-13278-PA. DOI: 10.2118/13278-PA.

Nolte, K.G. 1986b. A General Analysis of Fracturing Pressure Decline With Application to Three Models. *SPEPE* **1** (6): 571–583. SPE-12941-PA. DOI: 10.2118/12941-PA.

Nolte, K.G. 1988. Principles for Fracture Design Based on Pressure Analysis. *SPEPE* **3** (1): 22–30. SPE-10911-PA. DOI: 10.2118/10911-PA.

Nolte, K.G. and Smith, M.G. 1981. Interpretation of Fracturing Pressures. *JPT* **33** (9): 1767–1775. SPE-8297-PA. DOI: 10.2118/8297-PA.

Nordgren, R.P. 1972. Propagation of a Vertical Hydraulic Fracture. *SPEJ* **12** (4): 306–314; *Trans.*, AIME, **253**. SPE-3009-PA. DOI: 10.2118/3009-PA.

Odeh, A.S. and Jones, L.G. 1965. Pressure Drawdown Analysis, Variable-Rate Case. *JPT* **17** (8): 960–964; *Trans.*, AIME, **234**. SPE-1084-PA. DOI: 10.2118/1084-PA.

Peaceman, D.W. 1978. Interpretation of Well-Block Pressures in Numerical Reservoir Simulation. *SPEJ* **18** (3): 183–194; *Trans.*, AIME, **265**. SPE-6893-PA. DOI: 10.2118/6893-PA.

Pearson, C.M., Bond, A.J., Eck, M.E., and Schmidt, J.H. 1992. Results of Stress-Oriented and Aligned Perforating in Fractured Deviated Wells. *JPT* **44** (1): 10–18; *Trans.*, AIME, **293**. SPE-22836-PA. DOI: 10.2118/22836-PA.

Penny, G.S. and Jin, L. 1995. The Development of Laboratory Correlations Showing the Impact of Multiphase Flow, Fluid, and Proppant Selection Upon Gas Well Productivity. Paper SPE 30494 presented at the SPE Annual Technical Conference and Exhibition, Dallas, 22–25 October. DOI: 10.2118/30494-MS.

Penny, G.S. and Jin, L. 1996a. Report on the Investigation of the Effects of Fracturing Fluids Upon the Conductivity of Proppants, Proppant Flowback and Leakoff. Preliminary annual report, Proppant Consortium, Stim-Lab, Duncan, Oklahoma (May 1996).

Penny, G.S. and Jin, L. 1996b. The Use of Inertial Force and Low Shear Viscosity To Predict Cleanup of Fracturing Fluids Within Proppant Packs. Paper SPE 31096 presented at the SPE Formation Damage Control Symposium, Lafayette, Louisiana, 14–15 February. DOI: 10.2118/31096-MS.

Penny, G.S. and Jin, L. 2004. Report on the Investigation of the Effects of Fracturing Fluids Upon the Conductivity of Proppants, Proppant Flowback, and Leakoff. Final report presented to the Stim-Lab Consortia, Mesa, Arizona, 26–27 February.

Perkins, T.K. and Kern, L.R. 1961. Widths of Hydraulic Fractures. *JPT* **13** (9): 937–949; *Trans.*, AIME, **222**. SPE-89-PA. DOI: 10.2118/89-PA.

Pope, D.S., Britt, L.K., Constein, V.G., Anderson, A., and Leung, L.K.-W. 1996. Field Study of Guar Removal From Hydraulic Fractures. Paper SPE 31094 presented at the SPE Formation Damage Control Symposium, Lafayette, Louisiana, 14–15 February. DOI: 10.2118/31094-MS.

Pope, D.S., Leung, L.K., Gulbis, J., Constein, V.G. 1996. Effects of Viscous Fingering on Fracture Conductivity. *SPEPF* **11** (4): 230–237. SPE-28511-PA. DOI: 10.2118/28511-PA.

Prats, M. 1961. Effect of Vertical Fractures on Reservoir Behavior—Incompressible Fluid Case. *SPEJ* **1** (2): 105–118; *Trans.*, AIME, **222**. SPE-1575-G. DOI: 10.2118/1575-G.

Raghavan, R. 1993. *Well Test Analysis*. Englewood Cliffs, New Jersey: Prentice-Hall.

Rawlins, E.L. and Schellhardt, M.A. 1935. *Back-Pressure Data on Natural Gas Wells and Their Application to Production Practices*. Monograph, US Bureau of Mines, Washington, DC **7**.

Robertson, E.C. 1964. Viscoelasticity of rocks. In *State of Stress in the Earth's Crust: Proceedings of the International Conference 1963*, ed. W. Judd, 181–224. Oxford, UK: Elsevier.

Rowley, D.S., Burk, C.A., and Manual, T. 1981. Oriented Cores. Technical Report, Christensen Diamond Products, Salt Lake City, Utah (February 1981).

RP 39, Recommended Practice for Standard Procedures for the Evaluation of Hydraulic Fracturing Fluids. 1983. Washington, DC: API.

RP 56, Recommended Practices for Testing Sand Used in Hydraulic Fracturing Operations. 1983. Washington, DC: API.

RP 60, Recommended Practices for Testing High Strength Proppants Used in Hydraulic Fracturing Operations. 1989. Washington, DC: API.

Samuelson, M.L. and Constein, V.G. 1996. Effects of High Temperature on Polymer Degradation and Cleanup. Paper SPE 36495 presented at the SPE Annual Technical Conference and Exhibition, Denver, 6–9 October. DOI: 10.2118/36495-MS.

Schlottman, B.W., Miller, W.K. II, and Leuders, R.K. 1981. Massive Hydraulic Fracture Design for the East Texas Cotton Valley Sands. Paper SPE 10133 presented at the SPE Annual Technical Conference and Exhibition, San Antonio, Texas, 4–7 October. DOI: 10.2118/10133-MS.

Schubarth, S. and Milton-Tayler, D. 2004. Investigating How Proppant Packs Change Under Stress. Paper SPE 90562 presented at the SPE Annual Technical Conference and Exhibition, Houston, 26–29 September. DOI: 10.2118/90562-MS.

Schulte, W.M. 1986. Production From a Fractured Well With Well Inflow Limited to Part of the Fracture Height. *SPEPE* **1** (5): 333–343. SPE-12882-PA. DOI: 10.2118/12882-PA.

Settari, A., Bale, A., Bachman, R.C., and Floisand, V. 2002. General Correlation for the Effect of Non-Darcy Flow on Productivity of Fractured Wells. Paper SPE 75715 presented at the SPE Gas Technology Symposium, Calgary, 30 April–2 May. DOI: 10.2118/75715-MS.

Shah, S.N., Lord, D.L., and Tan, H.C. 1992. Recent Advances in the Fluid Mechanics and Rheology of Fracturing Fluids. Paper SPE 22391 presented at the International Meeting on Petroleum Engineering, Beijing, 24–27 March. DOI: 10.2118/22391-MS.

Shelley, R.F. and McGowen, J.M. 1986. Pump-in Test Correlation Predicts Proppant Placement. Paper SPE 15151 presented at the SPE Rocky Mountain Regional Meeting, Billings, Montana, USA, 19-21 May 1986. DOI: 10.2118/15151-MS.

Simonson, E.R., Abou-Sayed, A.S., and Clifton, R.J. 1978. Containment of Massive Hydraulic Fractures. *SPEJ* **18** (1): 27–32. SPE-6089-PA. DOI: 10.2118/6089-PA.

Small, J., Wallace, M., Van Howe, S., Brown, E., Thompson, D., and Dowell, J. 1991. Improving Fracture Conductivities With a Delayed Breaker System: A Case History. Paper SPE 21497 presented at the SPE Gas Technology Symposium, Houston, 23–25 January. DOI: 10.2118/21497-MS.

Smith, M.B. 1985. Stimulation Design for Short, Precise Hydraulic Fractures. *SPEJ* **25** (3): 371–379. SPE-10313-PA. DOI: 10.2118/10313-PA.

Smith, M.B., Bale, A., Britt, L.K., Cunningham, L.E., Jones, J.R., Klein H.H., and Wiley, R.P. 2004. An Investigation of Non-Darcy Flow Effects on Hydraulic Fractured Oil and Gas Well Performance. Paper SPE 90864 presented at the SPE Annual Technical Conference and Exhibition, Houston, 26–29 September. DOI: 10.2118/90864-MS.

Smith, M.B., Bale, A.B., Britt, L.K., Klein, H.H., Siebrits, E., and Dang, X. 2001. Layered Modulus Effects on Fracture Propagation, Proppant Placement, and Fracture Modeling. Paper SPE 71654 presented at the SPE Annual Technical Conference and Exhibition, New Orleans, 30 September–3 October. DOI: 10.2118/71654-MS.

Smith, M.B., Miller, W.K. II, and Haga, J. 1987. Tip Screenout Fracturing: A Technique for Soft, Unstable Formations. *SPEFE* **2** (2): 95–103; *Trans.*, AIME, **283**. SPE-13273-PA. DOI: 10.2118/13273-PA.

Smith, M.B., Ren, N.-K., Sorrels, G.G., and Teufel, L.W. 1986. A Comprehensive Fracture Diagnostics Experiment. Part II—Comparison of Fracture Azimuth Measuring Procedures. *SPEPE* **1** (6): 423–431; *Trans.*, AIME, **281.** SPE-13894-PA. DOI: 10.2118/13894-PA.

Smith, M.B., Reeves, T.L., and Miller, W.K. II. 1989. Multiple Fracture Height Measurements: A Case History. Paper SPE 19092 presented at the SPE Gas Technology Symposium, Dallas, 7–9 June. DOI: 10.2118/19092-MS.

Smith, M.B., Rosenburg, R.J. and Bowen, J.F. 1982. Fracture Width—Design vs. Measurement. Paper SPE 10965 presented at the SPE Annual Technical Conference and Exhibition, New Orleans, 26–29 September. DOI: 10.2118/10965-MS.

Soliman, M.Y. and Hunt, J.L. 1985. Effect of Fracturing Fluid and Its Cleanup on Well Performance. Paper SPE 14514 presented at the SPE Eastern Regional Meeting, Morgantown, West Virginia, USA, 6–8 November. DOI: 10.2118/14514-MS.

Spath, J.B., Ozkan, E., and Raghavan, R. 1994. An Efficient Algorithm for Computation of Well Responses in Commingled Reservoirs. *SPEFE* **9** (2): 115–121. SPE-21550-PA. DOI: 10.2118/21550-PA.

Spivey, J.P. and Lee, W.J. 1999. Application of the Diagnostic Plot Using a Derivative Based on Shut-In Time. Paper SPE 56424 presented at SPE Annual Technical Conference and Exhibition, Houston, 3–6 October. DOI: 10.2118/56424-MS.

Stark, A. 1998. Investigation of the influence of turbulence on deliverability of gas wells. MEng thesis, University of Calgary, Alberta, Canada.

Swift, S.C. 1988. Application of Equivalent Drawdown Time in Well Testing. Paper SPE 17547 presented at the SPE Rocky Mountain Regional Meeting, Casper, Wyoming, 11–13 May. DOI: 10.2118/17547-MS.

Tarantola, A. 1987. *Inverse Problem Theory Methods for Fitting and Model Parameter Estimation.* New York: Elsevier.

Teufel, L.W. 1982. Prediction of Hydraulic Fracture Azimuth from Anelastic Strain Recovery Measurements of Oriented Core. *Proc.*, 23rd U.S. Symposium on Rock Mechanics, Berkeley, California, 238–246.

Teufel, L.W., Hart, C.M., Sattler, A.R., Clark, J.A., and Calvin, C. 1984. Determination of Hydraulic Fracture Azimuth by Geophysical, Geological, and Oriented-Core Methods at the Multiwell Experiment Site, Rifle, CO. Paper SPE 13226 presented at the SPE Annual Technical Conference and Exhibition, Houston, 16–19 September. DOI: 10.2118/13226-MS.

Theis, C.V. 1935. The relation between the lowering of the piezometric surface and the rate and duration of discharge of a well using ground-water storage. *American Geophysical Union Transactions* **16** (2): 519–524.

Thorpe, R. and Springer, J. 1982. Relationship between borehole elongation and in situ stress orientation at the Nevada Test Site. Paper presented at the US Rock Mechanics Symposium, Berkeley, California, 25–27 August.

Tinker, S.J., Baycroft, P.D., Ellis, R.C., and Fitzhugh, E. 1997. Mini-Frac Tests and Bottomhole Treating Pressure Analysis Improve Design and Execution of Fracture Stimulations. Paper SPE 37431 presented at the SPE Production Operations Symposium, Oklahoma City, Oklahoma, USA, 9–11 March. DOI: 10.2118/37431-MS.

van Everdingen, A.F. and Hurst, W. 1949. The Application of the Laplace Transformation to Flow Problems in Reservoirs. *Trans.,* AIME, **186:** 305–324.

Veatch, R.W. Jr. 1983a. Overview of Current Hydraulic Fracturing Design and Treatment Technology—Part 1. *JPT* **35** (4): 677–687. SPE-10039-PA. DOI: 10.2118/10039-PA.

Veatch, R.W. Jr. 1983b. Overview of Current Hydraulic Fracturing Design and Treatment Technology—Part 2. *JPT* **35** (5): 853–864. SPE-11922-PA. DOI: 10.2118/11922-PA.

Veatch, R.W. Jr. and Crowell, R.F. 1982. Joint Research/Operations Programs Accelerate Massive Hydraulic Fracturing Technology. *JPT* **34** (12): 2763–2775. SPE-9337-PA. DOI: 10.2118/9337-PA.

Voneiff, G.W., Robinson, B.M., and Holditch, S.A. 1996. The Effects of Unbroken Fracture Fluid on Gaswell Performance. *SPEPF* **11** (4): 223–229. SPE-26664-PA. DOI: 10.2118/26664-PA.

Warpinski, N.R. 1983. Investigation of the accuracy and reliability of in situ stress measurements using hydraulic fracturing in perforated, cased holes. *Proc.*, 24th U.S. Symposium on Rock Mechanics, College Station, Texas, 773–786.

Warpinski, N.R., Branagan, P., and Wilmer, R. 1985. In-Situ Stress Measurements at U.S. DOE's Multiwell Experiment Site, Mesaverde Group, Rifle, Colorado. *JPT* **37** (3): 527–536. SPE-12142-PA. DOI: 10.2118/12142-PA.

Warpinski, N.R. and Teufel, L.W. 1989. In-Situ Stresses in Low-Permeability, Nonmarine Rocks. *JPT* **41** (4): 405–414; *Trans.,* AIME, **287**. SPE-16402-PA. DOI: 10.2118/16402-PA.

Willberg, D.M., Card, R.J., Britt, L.K., Samuel, M., England, K.W., Cawiezel, K.E., and Krus, H. 1997. Determination of the Effect of Formation Water on Fracture Fluid Cleanup Through Field Testing in the East Texas Cotton Valley. Paper SPE 38620 presented at the SPE Annual Technical Conference and Exhibition, San Antonio, 5–8 October. DOI: 10.2118/38620-MS.

Wood, M.D. 1981. Method of determining change in the subsurface structure due to application of fluid pressure to the earth. US Patent No. 4,271,696.

Wood, M.D., Pollard, D.D., and Raleigh, C.B. 1976. Determination of In-Situ Geometry of Hydraulically Generated Fractures Using Tiltmeters. Paper SPE 6091 presented at the SPE Annual Technical Conference and Exhibition, New Orleans, 3–6 October. DOI: 10.2118/6091-MS.

Author Index

Subject Index

A

acid, 2, 22, 47, 51, 75, 92
additive bins, 9
Agarwal-Gardner type curves, 115–117, 125–127
Agarwal's pseudotime, 103–105
algorithmic time derivative
    bilinear flow and, 83, 85–86
    conductivity and, 86–87, 89–92
    diagnostic plots and, 84, 86–87, 100–102
    fracture height and, 89
    history matching and, 87–88
    interpretation and analysis algorithm and, 82–90
    linear flow and, 83
    permeability and, 86–87
    proppants and, 86
    single-point flow and buildup test and, 99–103
    skin factor and, 88–89
    transition zones and, 88–89
    vertical fractures and, 82–89
    wellbore storage and, 83, 85, 87–89
American Petroleum Institute (API), 25–26
ammonium persulfate, 22
Amoco Production Company, 1–2
analytical methods, 118–120
Archer County, Texas, 5
Arco, 10
artificial-lift systems, 112
azimuth determination, 31–32

B

Barnett shale, 43
bedding, 19
bilinear flow, 81, 92
    algorithmic time derivative and, 83, 85–86
    drawdown test interpretation and, 92–95
    pressure buildup testing and, 107, 109
Blasingame-McCray type curves, 115–117
bottomhole treating pressure (BHTP), 43–45, 47, 55, 61
boundary conditions
    Darcy's law and, 73
    numerical methods and, 121–122
    type curve methods and, 111–118, 125–127
boundary-dominated flow, 79–80, 83
BP, 10
breakers, 6, 22
    cleanup and, 23–24
    encapsulated, 23–24
    enzyme, 23
breakouts, 31–32

b

b-stem, 115
bubble sensors, 38

C

Canadian Rockies, 57
carbon dioxide expansion, 24
Carthage Gas Unit, 42
central difference formula, 122
Chase formation, 2
Cinco-Ley equivalent wellbore radius, 119
cleanup
    breakers and, 23–24
    effective conductivity and, 28–29
    performance prediction and, 123
completions cycle, 69–70
compressibility, 96
    boundary-dominated flow and, 79–80
    numerical methods and, 121–122
    pressure and, 103
    temperature and, 103
    wellbore storage and, 78–79
compressional velocity, 31
conductivity, 12, 81
    algorithmic time derivative and, 84–92
    cleanup and, 28–29
    dimensionless, 84–85, 89, 95
    drawdown test interpretation and, 92–95
    effective, 28–29
    equivalent wellbore radius and, 89–90
    finite, 90–91
    folds of increase (FOI) concept and, 13
    fracturing fluids and, 23
    infinite, 90–91
    numerical methods and, 121
    pressure buildup tests and, 123–124
    proppants and, 25
    tip-screenout (TSO) technique and, 26–29
connectivity, 81
    perfect, 72
    skin factor and, 75–78
    wellbore storage and, 78
constant terminal pressure, 113, 118–120
core analysis
    permeability and, 36
    Poisson's ratio and, 35
    post-appraisal and, 35
    prefrac data collection and, 35–37
    pretreatment planning and, 7–9
    triaxial compression test and, 35–36
    Young's modulus and, 35–36